Heating with Wolves, Cooling with Cacti

Thermo-bio-architectural Framework (ThBA)

Negin Imani

School of Architecture
Victoria University of Wellington
Wellington, New Zealand

Brenda Vale

School of Architecture
Victoria University of Wellington
Wellington, New Zealand

CRC Press
Taylor & Francis Group
Boca Raton   London   New York

CRC Press is an imprint of the
Taylor & Francis Group, an **informa** business

A SCIENCE PUBLISHERS BOOK

Cover illustration by Sarah Wilson of Wilson Ink Art (https://www.wilsoninkart.com/).

First edition published 2022
by CRC Press
6000 Broken Sound Parkway NW, Suite 300, Boca Raton, FL 33487-2742

and by CRC Press
4 Park Square, Milton Park, Abingdon, Oxon OX14 4RN

ISBN: 978-0-367-53438-7 (hbk)
ISBN: 978-0-367-53440-0 (pbk)
ISBN: 978-1-003-08193-7 (ebk)

DOI: 10.1201/9781003081937

Typeset in Palatino
by Radiant Productions

Preface

This book was developed from the PhD work of Negin Imani, which was partly supervised by Brenda Vale. However, this book is a substantial reworking of the thesis with additional chapters to make it accessible for many readers. Inspiration for the topic came from seeing how many modern buildings are poorly adapted to the climates and situations where they are located, unlike the local flora and fauna. Would it be possible to learn from how animals and plants thermo-regulate and find a way of transferring this knowledge to building designers? This was the underlying question that led to this book.

The big challenge came from trying to find a way to link two seemingly disparate bodies of knowledge—biology and architecture. The structure of the book to an extent follows the path taken by the research investigation. Having examined why there is a need for energy-efficient buildings in Chapter 1 and whether buildings have ever been better adapted to location in the way of the local flora and fauna in Chapter 2, Chapters 3 and 4 look at what has already been happening in the field of buildings that draw inspiration from nature, especially when it comes to energy efficiency. The starting point of any research is to build on what has gone before, just as this book will hopefully be a stepping stone for those with a fascination for bio-mimicry. The biggest challenge in developing the ThBA was to find a way of structuring existing information on how animals and plants thermo-regulate. This process is described in Chapter 5, and in turn structures how the types of thermo-regulation are presented in Chapter 6. The next step was to try and find a way of making this biological information both useful and accessible for designers. As a first step parallels were drawn between what happens in nature and what happens in buildings when it comes to thermo-regulation, as set out in Chapter 7. This preparatory knowledge led to the creation of the first version of the ThBA. We were privileged to have had the help of a panel of biologists from Victoria University of Wellington in evaluating this first version and we are eternally grateful for the time they gave to this project. The next step in the research was to test the ThBA to see if it were helpful for designers. This led to a number of revisions as described in Chapter 8, which also deals with using it for the first time to try and find solutions for the thermal challenges of two case study buildings in New Zealand.

What the research revealed, as discussed in the final chapter, is that building designers already use many of the thermo-regulatory strategies found in nature but not in the complex hierarchical way that they occur there. This is an area that yet remains to be explored.

Negin Imani
Brenda Vale

Contents

Chapter 1
Building Energy Use and Climate Change

1.1 Climate change

Global warming and consequent climate change is a universal issue that for many people is hard to appreciate as a problem because the link between everyday behaviour and its effect on climate is difficult to see. Only recently have a number of extreme weather events begun to be linked with the effect of changes to the climate, caused by increased concentrations of greenhouse gases (GHG) in the atmosphere. The problem in part stems from the fact that many buildings, especially those in the developed world are powered by energy that comes from fossil fuels, and it is the burning of these that has been linked to climate change.

Research has shown that global warming will continue to increase if appropriate measures and policies are not set in place and adhered to by influential industries (Chen 2015). Many countries have set out policies and initiated research for controlling and mitigating the potentially excruciating effects of climate change on human lives. An example is *CarbonWatchNZ*, a New Zealand (NZ) programme that focuses on the carbon balance of the whole country through measuring greenhouse gasses in the atmosphere (NIWA 2019). The NZ Ministry of Business, Innovation and Employment (MBIE 2019) states, "Approximately 20% of all energy in NZ is consumed in the operation of buildings and around 65–70% of the energy consumed in buildings is in the form of electricity. Buildings use slightly more than half of the electricity produced in NZ." Because NZ energy has a significant renewable component of total primary energy supply in 2019 (39.5%) (MBIE 2020) and 82.4% of electricity generated that same year (MBIE 2020), this does not seem a great problem. However, globally in 2015, "82% of final energy consumption in buildings was supplied by fossil fuels" (Abergel et al. 2017), and climate change is a global problem not just an NZ problem.

Climate change and global warming are expected to affect several aspects of building performance and this will happen in various ways. Extreme weather events leading to heavy rain could have an impact on building detailing and buildings may need to be placed or repositioned to avoid frequent flooding events. Increase in temperature will also affect building energy use and given buildings last a long time, this may be very different from the energy use predicted when the building was designed.

In the natural world, changes to the climate can lead to extinction, migration and adaptation (if remaining in the same place) of various species. For example, in Mexico climate change has already been linked to the extinction of some species of lizard. "We found a correlation between rate of change in T_{max} during winter-spring breeding periods and local extinctions of *Sceloporus* species" (Sinervo et al. 2010). Climate change is also leading to migration as "A tally of more than 4,000 species from around the world shows that roughly half are on the move" (Welch 2017). It is known that in warmer years, malaria-carrying mosquitoes migrate to higher altitudes, suggesting that "... climate change will, without mitigation, result in an increase of the malaria burden in the densely populated highlands of Africa and South America" (Siraj et al. 2014). When it comes to adaptation, the human species should have no problem surviving since it has already learned to live in most places in the world with the exception of the extreme polar regions. Jenkins et al. (2015) also point out that the slow rate of change to buildings and energy supply infrastructure "...will mean that both building designers and those involved with energy provision have time to respond, even within the context of changing climate and building technology." However, it might still be prudent to design buildings that minimise energy use both now and in the future as part of climate change mitigation. The aim, after all, is to sustain the human way of life.

1.2 Sustainability and climate change

Sustainability is a broad concept that is hard to define but as suggested at the end of the last section, the aim is generally for people to sustain a way of living. This necessitates recognising that humanity lives on the resources of a finite planet with the benefit of external solar energy from the sun. However, this only deals with the environment within which humanity dwells. When people talk about sustainability, they often refer to its three dimensions, these being the economic, social and environmental, often called the three pillars of sustainability (Ortiz et al. 2009). As a result, sustainable design in architecture has been defined as the way quality of life can be improved through the synergistic relationships between the three pillars of sustainability. In its turn, building performance can also be evaluated from three perspectives: (1) the requirements of the occupants, (2) economic sense and (3) environmental performance (Leaman et al. 2010). The latter (3) involves evaluation of material and energy flows emanating from the characteristics of buildings (Lützkendorf et al. 2005). Resource conservation can be categorised into energy, material, water, and land conservation, of which energy conservation has been regarded as the most important issue affecting the environment (Akadiri et al. 2012). Reduction in energy consumption is also critical as buildings and the construction sector consume nearly 40% of total global energy consumption, a consumption that is rising rather than falling (GlobalABC 2019).

1.2.1 Ecologically Sustainable Design (ESD)

Ecologically Sustainable Design (ESD) emerged as the way of applying the 1990 Australian concept of Ecologically Sustainable Development to buildings (Gamage and Hyde 2012). The aim was always to integrate the environmental aspects of development with the economic aspects. As an approach towards the development of a sustainable built environment, ESD mainly targets energy consumption reduction (GhaffarianHoseini 2012), since reducing energy demand tends to make economic sense over the life of the building. This has made building energy efficiency the most prominent concept in ESD (Jabareen 2008). As part of being environmentally sustainable, a building needs to minimise the fossil fuel energy used for its construction, operation, and maintenance. Ideally, ESD also needs to minimise the land needed to supply its material resources, whether these be timber, masonry materials, metals or plastics, and the energy bound up in producing these materials. This means an environmentally sustainable building is expected to have low or zero environmental impact.

The reflection of this can also be seen in the different methods developed since early 1990s for assessing the environmental performance of buildings. While there is a lack of agreement on the full list of necessary environmental indicators (Passer et al. 2012), the Life Cycle Assessment (LCA) method seems to have been accepted as useful for evaluating the environmental impacts of a building. The LCA methodology complies with the International Standardisation Organisation (ISO) 14040 standards provided in 1997 to address different aspects of sustainable building design and construction (International Organization for Standardization 2006).

An LCA analysis considers all stages of a building's life from production to use, and disposal. Each of these phases requires raw materials and energy as inputs, meaning that every stage of a building's life potentially emits carbon, and also creates waterborne and solid waste products. Among the environmental indicators, carbon emissions have been recognised as the most important because of their contribution to global warming. The evidence shows that the pace of global warming will shortly exceed that of the worst-case scenarios predicted for 2003 and 2005 (Roaf et al. 2009). The effect of human-induced global warming has contributed to recent heavy precipitation events (Min et al. 2011). Given that greenhouse gas emissions are one of the factors that affect climate change (Carter et al. 2015), and as noted above, given that buildings are huge consumers of energy (Stojiljković et al. 2015), the design of energy-efficient buildings should make a substantial difference in climate change mitigation.

Life Cycle Energy Analysis (LCEA) exclusively evaluates the total amount of energy inputs to a building during its life span. LCEA is thus a derivative of LCA but only deals with energy, and by implication, carbon, if that energy comes from fossil fuels. Calculation of the embodied energy (EE) and operational energy (OE) indicates the total energy a building consumes within its lifetime (Fay et al. 2000).

Embodied energy is the energy consumed for the mining and production of raw materials and transportation of materials to the site. In a building, the embodied energy comes from the materials used both in its initial construction and in its maintenance. The energy required for maintaining comfort conditions is known as the operational energy, which is used for running HVAC systems, providing hot water, lighting and running other appliances. In addition to EE and OE, some researchers have added the third category of demolition energy (DE), which relates to the energy used for the demolition and recycling phase of a building's lifespan (Cabeza et al. 2014). However, DE is often ignored because it is very small compared to EE and OE (Fay et al. 2000). Both the construction and use of buildings lead to significant negative environmental impacts which also normally contribute to climate change.

1.3 Biomimicry

Biomimicry has been proposed as a means of merging environmental consideration into design projects in order to achieve sustainability (Wahl 2006). Biomimicry has also been associated with innovation, which might or might not lead to building sustainability (Section 3.1.2 in Chapter 3). The question is whether biomimicry has much to offer energy-efficient design and this is explored in this book. Technological innovations have been shown to increase the energy efficiency of energy-consuming systems (Herring and Roy 2007), and in the same context, biomimicry has been recognised as an innovative design approach for improving energy-efficient design (Lurie-Luke 2014, Radwan and Osama 2016, Pedersen Zari 2018). As Angela Nahikian stated: "Nature is constantly innovating, endlessly experimenting and ever reinventing itself in the face of new challenges. From materials and products to business models, biomimicry offers a fresh lens for all the dreamers and doers remaking the man-made world" (The Biomimicry Institute 2019). It has also been suggested that biomimetic design might provide opportunities for creating a shift in the way design normally proceeds (Vincent et al. 2005).

One of the benefits of adopting biomimicry principles in the construction industry is the potential reduction in global warming. It seems that nature uses low-energy processes (Oguntona and Aigbavboa 2018) and this suggests the presence of numerous examples of biological organisms which could be explored for the energy-efficient processes they use to inform innovative solutions for solving human design problems. The purpose of this book is to seek out these examples and find a way of presenting them that makes it easy for designers to look for energy-efficient solutions in nature that can be applied to buildings.

References

Abergel, T., Dean, B. and John Dulac, J. (2017). *Global Status Report 2017, United Nations Environment/IEA*. Retrieved from https://www.worldgbc.org/sites/default/files/UNEP%20188_GABC_en%20%28web%29.pdf.

Akadiri, P. O., Chinyio, E. A. and Olomolaiye, P. O. (2012). Design of a sustainable building: A conceptual framework for implementing sustainability in the building sector. *Buildings, 2*(2), 126–152.

Cabeza, L. F., Rincón, L., Vilariño, V., Pérez, G. and Castell, A. (2014). Life cycle assessment (LCA) and life cycle energy analysis (LCEA) of buildings and the building sector: A review. *Renewable and Sustainable Energy Reviews, 29*, 394–416.

Carter, J. G., Cavan, G., Connelly, A., Guy, S., Handley, J. and Kazmierczak, A. (2015). Climate change and the city: Building capacity for urban adaptation. *Progress in Planning, 95*, 1–66.

Chen, Y. (2015). *Environmental Adaptive Design Building Performance Analysis Considering Change* (Master's). University of South Carolina, Colombia, SC (10799532).

Fay, R., Treloar, G. and Iyer-Raniga, U. (2000). Life-cycle energy analysis of buildings: a case study. *Building Research & Information, 28*(1), 31–41.

Gamage, A. and Hyde, R. (2012). A model based on Biomimicry to enhance ecologically sustainable design. *Architectural Science Review, 55*(3), 224–235. doi:10.1080/00038628.2012.709406.

GhaffarianHoseini, A. (2012). Ecologically sustainable design (ESD): theories, implementations and challenges towards intelligent building design development. *Intelligent Buildings International, 4*(1), 34–48.

GlobalABC. (2019). *2019 Global Status Report for Buildings and Construction: Towards A Zero-Emission, Efficient and Resilient Buildings and Construction Sector.* Retrieved from https://wedocs.unep.org/bitstream/handle/20.500.11822/30950/2019GSR.pdf, Accessed 21 September 2020.

Herring, H. and Roy, R. (2007). Technological innovation, energy efficient design and the rebound effect. *Technovation, 27*(4), 194–203.

International Organization for Standardization. (2006). Environmental Management: Life Cycle Assessment; Principles and Framework. *In*: ISO.

Jabareen, Y. (2008). A new conceptual framework for sustainable development. *Environment, Development and Sustainability, 10*(2), 179–192.

Jenkins, D. P., Patidar, S. and Simpson, S. A. (2015). Quantifying change in buildings in a future climate and their effect on energy systems. *Buildings, 5*(3), 985–1002.

Leaman, A., Stevenson, F. and Bordass, B. (2010). Building evaluation: practice and principles. *Building Research & Information, 38*(5), 564–577.

Lurie-Luke, E. (2014). Product and technology innovation: What can biomimicry inspire? *Biotechnology Advances, 32*(8), 1494–1505.

Lützkendorf, T., Speer, T., Szigeti, F., Davis, G., le Roux, P. C., Kato, A. and Tsunekawa, K. (2005). A comparison of international classifications for performance requirements and building performance categories used in evaluation methods. *In*: P. Huovila (ed.). *Performance based Building*. Espoo, Finland: Technical Research Centre of Finland.

MBIE. (2019). *Transforming Operational Efficiency*. Ministry of Business, Innovation and Employment Retrieved from https://www.mbie.govt.nz/dmsdocument/11793-transforming-operational-efficiency.

MBIE. (2020). *Energy in New Zealand 2020*. Ministry of Business, Innovation and Employment Retrieved from https://www.mbie.govt.nz/dmsdocument/11679-energy-in-new-zealand-2020.

Min, S.-K., Zhang, X., Zwiers, F. W. and Hegerl, G. C. (2011). Human contribution to more-intense precipitation extremes. *Nature, 470*(7334), 378.

NIWA. (2019). *A Bird's-Eye View of Our Carbon Balance*. Retrieved 20 June 2019, from NIWA https://www.niwa.co.nz/news/a-birds-eye-view-of-our-carbon-balance.

Oguntona, O. A. and Aigbavboa, C. O. (2018). *Benefits of Biomimicry Adoption and Implementation in the Construction Industry*. Paper presented at the International Conference on Applied Human Factors and Ergonomics, Orlando, FL.

Ortiz, O., Castells, F. and Sonnemann, G. (2009). Sustainability in the construction industry: A review of recent developments based on LCA. *Construction and Building Materials, 23*(1), 28–39. doi:10.1016/j.conbuildmat.2007.11.012.

Passer, A., Kreiner, H. and Maydl, P. (2012). Assessment of the environmental performance of buildings: A critical evaluation of the influence of technical building equipment on residential buildings. *The International Journal of Life Cycle Assessment, 17*(9), 1116–1130.

Pedersen Zari, M. (2018). *Regenerative Urban Design and Ecosystem Biomimicry*. London, England: Routledge.

Radwan, G. A. and Osama, N. (2016). Biomimicry, an approach, for energy efficient building skin design. *Procedia Environmental Sciences, 34*, 178–189.

Roaf, S., Crichton, D. and Nicol, F. (2009). *Adapting Buildings and Cities for Climate Change: A 21st Century Survival Guide* (2nd ed.). Burlington, MA: Architectural Press, Elsevier.

Sinervo, B., Méndez-de-la-Cruz, F., Miles, D. B., Heulin, B., Bastiaans, E., Villagrán-Santa Cruz, M., Lara-Resendiz, R., Martínez-Méndez, N., Calderón-Espinosa, M. L., Meza-Lázaro, R. N., Gadsden, H., Avila, L. J., Morando, M., De la Riva, I. J., Sepulveda, P. V., Rocha, C. F., Ibargüengoytía, N., Puntriano, C. A., Massot, M., Lepetz, V., Oksanen, T. A., Chapple, D. G., Bauer, A. M., Branch, W. R., Clobert, J. and Sites Jr., J. W. (2010). Erosion of lizard diversity by climate change and altered thermal niches. *Science, 328*(5980), 894–899.

Siraj, A., Santos-Vega, M., Bouma, M., Yadeta, D., Carrascal, D. R. and Pascual, M. (2014). Altitudinal changes in malaria incidence in highlands of Ethiopia and Colombia. *Science, 343*(6175), 1154–1158.

Stojiljković, M. M., Ignjatović, M. G. and Vučković, G. D. (2015). Greenhouse gases emission assessment in residential sector through buildings simulations and operation optimization. *Energy, 92*, 420–434.

The Biomimicry Institute. (2019). *History of the Biomimicry Institute*. Retrieved 29 July 2019 https://biomimicry.org/history/.

Vincent, J. F., Bogatyreva, O., Pahl, A.-K., Bogatyrev, N. and Bowyer, A. (2005). Putting biology into Triz: a database of biological effects. *Creativity and Innovation Management, 14*(1), 66–72. doi:10.1111/j.1476-8691.2005.00326.x.

Wahl, D. C. (2006). Bionics vs. Biomimicry: From control of nature to sustainable participation in nature. pp. 289–298. *In*: C. A. Brebbia (ed.). *Comparing Design in Nature with Science and Engineering* (Vol. 87). Southampton, UK: WIT.

Welch, C. (2017). *Half of All Species Are on the Move—And We're Feeling It*. Retrieved from https://www.nationalgeographic.com/science/article/climate-change-species-migration-disease.

Chapter 2
Thermal Issues and Building Design

2.1 Design and climate

The idea of biomimicry and finding inspiration from nature to solve problems is relatively new, but the human species has learned to live in a wide variety of environments and in an equally wide variety of buildings adapted to provide a degree of comfort in those same environments. This suggests that in the past, people had an understanding of their immediate environment that was gained through observation of the natural world. The approach of designing to suit climate—bioclimatic design—has been well described and documented (DeKay and Brown 2013, Hausladen 2005, Hyde 2000, Olgyay 2015). Traditional or vernacular buildings have evolved as a response to climate and also to the materials available to human builders. Some use will have been made of local stone but for materials that grow, such as timber, what was available to the vernacular builders would also have been affected by climate. Thus, timbers which are to an extent resistant to rot, such as tropical hardwoods like lignumvitae and old growth teak have developed in that way precisely because the warm temperatures and high moisture levels of a tropical climate are more conducive to rotting wood.

Before the ready availability of fossil fuels first to heat and then to cool buildings through the generation of electricity, design had to work with the local climate. Thus, the sun-dried adobe brick buildings of Egypt are noted because the Nile silt provided an appropriate material and the sun was available to dry the bricks as the shortage of timber meant there was little wood to fire bricks. The scarce timber available was needed for other purposes such as roofs, which were "...made with strips of wood and palm-branches woven together, the surface being plastered with mud" (Polkinghorne and Polkinghorne 1949). In thermal terms, such buildings were ideal for a hot dry climate where the days are hot but the nights are much colder. Because the walls of mud brick buildings are thick for structural reasons as the mud brick has a low bearing capacity, the sun shining on the walls heats them from the outside with the heat travelling inwards, arriving in the early evening when the outside temperature starts to drop. Until this heat pulse arrives, the inside of the mud brick building is much cooler than the outside. As the outside temperature drops and while people sleep, the heat starts flowing from the inside out so the wall cools and is ready to be heated again in the next daily cycle. This is a prime example of bioclimatic design.

The question of importance for this book is not local materials and how they have been used to provide human comfort in dwellings, as in the example of the mud brick buildings, but whether the way people live in different climates is reflected in the types and behaviour of the local flora and fauna. As an example, does the way desert animals keep both cool and warm reflect the human example of using mud brick as a thermal regulator? The body temperature of a camel does increase during the day and drop at night, so at first sight there is a similarity but the camel's ability to thermoregulate is also affected by whether water is restricted and how much food is eaten (Bouâouda et al. 2014). The thermoregulation of the body of the camel is thus much more complex than the flow of heat through a mud brick wall. The latter is more a case of helping the thermoregulation of human body temperature by reducing temperature extremes inside a building.

To investigate the link between design in a particular climate and thermoregulation in flora and fauna in a similar environment, a number of brief case studies of vernacular buildings are discussed below, picked to reflect a variety of different climates.

2.1.1 Cold winters, cool summers

Traditional houses in the forested areas of northern Europe—Finland, Sweden and Russia—were built of notched logs, with cracks stuffed with moss to keep out the wind. Buildings were typically small and low with tiny openings and with a low-pitched roof covered with sod that was capable of shedding rain but able to withstand the winter storms (Phleps 1982). Covered with a thick layer of snow the low-pitched roof was even better insulated. The plentiful supply of wood meant the building could be heated in the very cold winter but keeping it small helped to minimise the work of preparing and stacking firewood. In the warmer summers, work could be taken outside but no changes were made to the building apart from opening doors and windows. Perhaps the first thing to note is just as the coniferous trees of these forests do not appear to change throughout the year, though they do over time shed needles and grow new ones, so the vernacular houses do not change. However, the trees are alive whereas most of the house is not, given the sod roof may flower in summer.

When it comes to the Finnish reindeer, which might live in such forest: "Energy expenditure is lower in winter than in summer ... Heat is saved by the effective fur insulation and peripheral heterothermia [a heterothermic animal varies between self-regulating its body temperature and allowing it to be affected by the surrounding environment] of the extremities" (Soppela et al. 1986). The latter appears to occur as a result of countercurrent heat exchange in the legs of the reindeer between arteries and veins, together with some reduction in blood flow at very low temperatures (Olsson 2011). Although insulation plays a part in keeping the reindeer at an acceptable temperature, the traditional log structure of the human-made house is not particularly well insulated and relies on burning rather than conserving energy. In this vernacular example there is nothing similar to the countercurrent heat exchange found in reindeer legs, although keeping the building small and low avoids having cold extremities, something not possible for an animal that needs to move to find food.

The Eurasian Brown Bear (*ursus arctos*) is the national animal of Finland. It is also on the International Union for Conservation of Nature (IUCN) Red List of threatened species, although the current population is stable (McLellan et al. 2017). The Brown Bear uses the thermoregulation strategy of hibernation from late October to April in very cold areas such as northern Finland and parts of Siberia (BBC Wildlife 2020). In a study of *ursus arctos* in North America, Schwartz et al. (2003) state body temperature drops 4–5°C when a bear is hibernating, from a normal body temperature range of 36.5°C–38.5°C. The heart rate also drops from a resting rate of 40–50 beats/min to 8–12 beats/min. Although people living in similar cold climates might go out less in the very cold weather, there is no human equivalent of hibernation.

Plants have also had to adapt to cold climates and very short growing seasons. In the sub-arctic tundra where there are no trees, plants tend to be low to the ground and clustered together to reduce the effect of wind-borne ice particles (Wheeling Jesuit University 2004). The only parallel with human settlement might be the fact that vernacular houses tend to be low in profile and also clustered in small settlements. Other plants only flower in the short summer season so the seeds lie dormant, which is proving an issue as global warming leads to warmer winters reducing snow cover so that plant dormancy is interrupted "...often resulting in high mortality due to drought and frost stress" (Bokhorst et al. 2018). However, dormancy, like hibernation, is not a strategy used by people and their buildings.

2.1.2 Hot summers, cold winters

The warm days and colder nights of a desert climate and the performance of mud brick walls have already been discussed. However, wide seasonal temperature swings are an additional feature of some climates. As an example, Yazd, a city near the centre of Iran, has hot summers and much colder dry winters. July is the warmest month when the average maximum temperature is 39°C, compared to an average maximum of 12°C for the coldest month of January. The average minimum temperature in January is –1°C (Weather and Climate n.d.). To cope with this swing in climate, a traditional house based on the courtyard evolved. "Houses are mostly single-storey and are densely clustered, side to side, back to back, along narrow streets, which are flanked by high walls with few openings"(Memarian and Brown 2003). The houses were built of adobe brick with very thick walls (Khajehzadeh et al. 2016), so these could absorb solar energy in the summer with the clustering ensuring the walls shaded the open areas such as the narrow streets. However, another strategy was also employed in the traditional courtyard houses, that of migrating between different parts of the house in summer and winter. The low winter sun would warm the rooms facing south in winter, while rooms facing north that were always in the shade would be preferred in summer. A survey undertaken in 2008 in Yazd found 69% of those living in traditional courtyard houses had rooms they used in different seasons (Foruzanmehr 2016). Traditional houses also had underground rooms which would be used at noon in summer (Khajehzadeh et al. 2016).

The other important vernacular features of houses in Yazd are the windcatchers or Badgir, which form a cooling mechanism. These chimney structures are higher than the house roof to catch the wind. Around Yazd, where the prevailing wind can be hot and dusty, the vents at the top of the wind catcher face away from the prevailing wind direction. In this case, they draw air up from the house. The air comes from the courtyard where there is normally a fountain, and this cooled air is then drawn through the summer rooms before venting up the windcatcher (Roaf 2011).

The sand cat (*Felis margarita*) is a desert-dwelling animal that has been recorded in Yazd Province, and is a mammal that has adapted to living in a desert climate. The sand cats have thick fur, especially on the feet, and to avoid the high heat in the middle of the day they dig burrows with their blunted claws that seem to be adapted for digging (Ghadirian et al. 2016). Unlike humans with their clustered houses to keep out the sun and wind, the sand cats are solitary animals that only come together for mating, needing to have sufficient space to prey on the rodents that live on the sparse vegetation that form an important part of their diet. They will drink water but can live on metabolic water (Ghadirian et al. 2016). Using underground space as a means of avoiding the hottest sun at midday is similar to using underground rooms at midday in traditional courtyard houses. The sand cats share the burrows but there is only ever one cat in occupation at a time (Ghadirian et al. 2016).

Plants that grow in desert regions have to be resistant to heat and often to high soil salinity. The tamarisk (known as salt cedar in America) is found in the desert around Yazd and has been described as being a plant that is especially drought resistant and adapted to saline soils (Carleton 1914). *Tamarix chinensis* has the ability to "...draw moisture from the saturated zones below the water table but may also survive indefinitely in the absence of saturated soil" (Everitt 1980). The leaves are fine and feathery in appearance, which helps to protect them from solar exposure. Despite its hardiness, the adaptation of tamarisk to a desert environment seems to have little in common with the adaptation found in the traditional courtyard houses of Yazd.

2.1.3 Hot, wet climates

Hot, wet climates are associated with the tropics. In tropical climates, there are no distinct thermal seasons as found in more temperate climates. Tropical regions, usually defined as lying between the Tropics of Cancer and Capricorn, are also distinguished by having a surplus of solar radiation in terms of gains and losses (Richter 2014). The rainy seasons in the tropics vary from place to place and Richter (2014) points out that the type of plants in a location indicates the extent of wet and humid months. For example, "...seasonal rainforests comprising less than 20% of deciduous trees are characteristic for zones of 1–3 arid months..." and "...deciduous tropical forests identify those with 6–8 arid months." In the hot, humid areas, traditional houses were elevated from the ground so that air could flow underneath. Roofs had large overhangs to provide shade and openings in the walls also allowed air flow through the building. A double roof was often used with a gap between the lower pitched part and the higher pitched part (see Figure 2-1) to allow for air circulation. The higher pitch also gave somewhere for the hot air to collect, all with the aim of keeping floor level as cool as possible. The materials would be timber or bamboo for the structure and leaves for thatching from the immediate environment.

Snakes which are found in the tropics also keep cool by seeking shade, just as the roof shades the interior of the traditional house. In contrast, the red jungle fowl (*gallus gallus*), the main ancestor of the modern domestic chicken (Lawal et al. 2018), which are found in moist tropical forest areas from eastern India to Indonesia and the Philippines (Animalia 2018), use a different method as part of keeping cool. Their featherless wattles and combs allow the blood to come to the surface and so lose heat, a feature which has descended to the modern chicken where it may be less

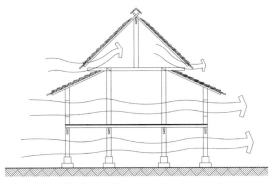

Figure 2-1. Traditional house section for wet, hot areas.

useful. There is no equivalent of this method in traditional houses in hot, moist climates. Closer to people, orangutans who live in the treetops of the rainforests of Borneo and Sumatra, and thus in the shade, will fan themselves with branches to keep cool (Shumaker et al. 2011), while the human history of the fan dates back to 3000 BC, with such fans being used both for cooling and in ceremonies (The Fan Museum 2020). Like the fan of the orangutan, early human fans were made of natural materials like leaves (Parr 2020).

2.1.4 *Traditional building*

The examples above show the dependence of traditional buildings on the availability of local materials, and thus there is a relationship between the plants in a particular location and the buildings there. Any link between the thermoregulatory strategies used by the local animals and the way the local people constructed those same buildings is much less clear. Vernacular building forms have been shown as working with climate to induce the most comfortable conditions, whether this is through keeping the building small and low with minimal openings in a cold climate or opening it up to the wind and elevating it above the ground in a hot dry climate. However, these strategies are much simpler than those used by some animals living in the same climate, from the heat recovery system of the legs of reindeer, to the 'cooling fins' of the wattles and comb of a red jungle fowl. This suggests that there may be thermoregulatory strategies still to be explored in a modern attempt to make buildings that are much more frugal in their use of energy, especially fossil fuel energy, than are being used at present.

2.2 Modern buildings

The vernacular approach to building using local materials began to be less possible as people started to live in cities, meaning such traditions are much more associated with rural than urban living. Some traditions, such as the courtyard houses of Yazd could be used in urban situations since their clustering was part of designing with climate, but as cities grew new housing forms were invented, such as the apartment blocks of Ostia, the sea port of Ancient Rome. Rome was also known for the height of its buildings. The fact "...the height of the city's buildings continued to awe the traveler for centuries after Rome had ceased to have any political importance is attested by the work of the late sixth century chronicler Zacharias" (Packer 1964). European medieval cities adopted the row or terraced house as a new housing form (Storey 2002), again suggesting that urbanisation tended to move buildings away from rural vernacular traditions. Living in cities also meant that resources had to be brought in from elsewhere, including the fuel needed for heating and cooking. This led to new problems: "As long ago as the year 1661, and probably long before that date, there were complaints about the pollution of London atmosphere by the fog or smoke of sea coal" (Anon 1902). The problem was made worse as in European cities and those established elsewhere on the European model, where urbanisation was associated with industrialisation as the factories spawned the cities (Gollin et al. 2015), the burning of fuels increased to keep the wheels of industry turning. A study of lake sediments in Southeast Asia has shown that industrialisation and "...the onset of extensive and region-wide fossil fuel consumption" was linked to evidence of air pollution (Engels et al. 2018). These urban pollution issues were one impetus behind the idea of creating more energy-efficient buildings to reduce fossil fuel use and the subsequent pollution.

Following the United Nations conference on the environment in Stockholm in 1972, there was a brief upsurge of interest in environmental matters such as pollution and resource depletion. The idea that oil might not last for ever was given further popular emphasis as a result of the OPEC oil embargo in the mid-1970s, when motorists suddenly could not obtain all the fuel they wanted (Vale and Vale 1991).

The effect on buildings, especially houses, of this situation was a series of experiments in designing to reduce the energy used in operating the building. This also brought back the idea of designing with climate in mind, as different techniques were suitable for different parts of the world. Design was also split between passive systems, which used the building fabric and orientation to achieve tolerable indoor conditions, and those that used active systems to circulate heated air or water around the building. The key to both systems in most climates was ensuring appropriate levels of insulation in the building fabric as a means of slowing heat loss and gain.

2.2.1 Passive solar design

Passive solar design was hardly a new concept in the 1970s since the Ancient Greeks had made use of it in their houses and cities (Butti and Perlin 1980) and the Ancient Romans in both their houses and bathhouses (Butti and Perlin 1980). Passive solar design had also been exploited by the Keck brothers with their 1940s houses in USA (Barber 2016). The aim with all such designs was simply to face the house towards the sun, with large windows and smaller or even no openings on the opposite side of the house. With internal exposed masonry finishes, such as plastered brick or tiles floors, and an insulation layer between these and the external wall material, the aim was to store the heat in the building fabric for the times when the sun was not shining, such as at night or even for those days without bright sunshine. The animal equivalent would be a cat moving into the sun to warm and retreating to a shaded spot when it feels too hot. Since a building cannot move, the fabric plus the solar gain is trying to achieve reasonable comfort over time. The essential point is that neither the cat or the building change. A more recent example of passive solar design is the row of BedZed houses in south London (Figure 2-2).

Passive solar conversion can also be achieved by attaching a glazed conservatory on the sun-facing side of an existing house and opening the house to this space on sunny days. The big problem with passive design is potential overheating in summer unless shading of the large glass areas can be achieved, perhaps through planting of deciduous trees to allow the sunlight through in winter but block it in summer, or through good stack ventilation that vents out the unwanted hot air at the top of the glass. Problems may still arise through overheating in summer, which can lead to energy use for cooling and also when people start to heat the additional conservatory in winter, in search of more living space (Marsh 2017). If

Figure 2-2. Highly glazed south-facing facades at BedZed (Chance 2007).

people were prepared to put up with a wider range of temperatures, as animals are forced to do, these issues could be mitigated.

2.2.2 Active solar design

The essence of active solar design is that energy is collected in one place and then transferred to another, whether this is to some form of heat storage for later use or to an interior space to warm it. Slight changes may occur within the building in terms of opening and closing valves or running fans to blow heated air around. This has a parallel with the blood being pumped through the wattles and comb to cool a jungle fowl or domestic chicken by losing heat to the surrounding air. The blood vessels in the comb and wattles will dilate for maximum blood flow, so there has been a slight change in the bird. The problem with active solar design in buildings is that the energy to run the systems has to be set against the energy saved to make sure there is an energy gain. An example of an active solar house is a house in Wadestown, Wellington (Figure 2-3) that has a rock store and fan to blow hot air through it from the store to the house as required in winter and also for taking heat from the house in summer to avoid overheating. The house also has sensors and automatic opening windows (Aouni Architecture 2016).

Figure 2-3. Active solar house, Wadestown, New Zealand (Aonui Architecture n.d.).

2.2.3 Passive and active comparisons

In some ways it is harder to distinguish between passive and active buildings than it is between animals that use passive strategies where the animal itself does not change and active strategies where the body does change, such as in vasodilation to lose heat. This is because a passive building may change if blinds are pulled to help prevent solar glare or doors and windows are opened to vent excess hot air. What is common to both passive and active solar design is the use of appropriate levels of insulation so as to minimise the energy needed to heat the house in winter so that the solar gain has a chance of making an effective contribution. Insulation in this instance is analogous to fur and feathers but with the exception that insulation tends to be fixed in the building envelope while feathers can be fluffed up in cold weather for increased insulation (Stettenheim 2006) and fur thins in the summer through moulting and thickens in the winter through the growth of underfur (Beltran et al. 2018).

Insulation is the common means that modern buildings use both to control heat loss in cold climates and to control heat gain in hot climates. That said, many modern buildings, especially commercial buildings have all glass exteriors which are not conducive to energy saving but rather tend to increase energy use for heating and cooling. Following the modernist ideal of the all glass tower in hot countries such as Egypt can lead to a position where the "...energy needed to cool the internal environment of these buildings is very high, leading to pressure on the total energy resources and the appearance of glass blocks that do not respect the architectural identity of the country" (Ahmed and Fikry 2019). This has led to designs for high-rise buildings in hot climates that attempt to use passive means to reduce energy use, such as the designs of Ken Yeang in Malaysia. An early example is the 1992 Mesiniaga Tower where the elevator core was placed to block the low-level western sun. The building has sunshades and also incorporates Yeang's signature facades and sky gardens (Yeang 2002). These buildings are examples of working with climate to make low-energy buildings rather than biomimicry.

2.3 Zero energy buildings

As part of climate change mitigation, the idea of the zero energy building emerged. Essentially, this is a very energy-efficient building that is linked to the grid and has some form of on-site generation whereby over a year it should be able to generate at least as much energy as it uses (Reeder 2016). As such it had similarities with the earlier autonomous and autarkic houses of the 1970s that were designed to collect all the energy needed from the site or plot where each house was situated, although without being linked to the grid. Autonomy went further by suggesting that the plot would also deal with a water supply and treat the wastes from those living in the house (Vale and Vale 1975). The problem with a zero energy building is drawing the boundary around what needs to be considered: Should the assessment include the energy that goes into making the building and into the energy-generating systems, whether these are photovoltaic panels (PVs) or small-scale wind generators? Since the building is grid linked, should the assessment include a share of the energy that goes into making the grid and the generating equipment? Even if the building can generate as much energy as it needs to operate over a year from renewable sources, such as the sun or wind, it is very improbable that all the energy to make its components will have been generated using renewable energy, at least in the near future.

Occupant behaviour can also have an effect on the performance of both zero energy and low-energy buildings (Stazi and Naspi 2018). This is where the parallels with the natural world break down. In nature, an animal or plant has to use its thermoregulatory systems to survive in its habitat but if the environmental parameters of the habit become too severe so that the animal or plant can no longer adapt, it will die, or move to another environment if it can. When it comes to zero-energy buildings there is always the option of drawing power from the gird to maintain human comfort, and this power may in most cases still be generated using fossil fuels. Although biomimicry encourages looking for solutions to human problems in the natural world, many people and certainly the majority in the developed world are currently not behaving in a changing climate situation so as to ensure survival of the human species in the same way that natural organisms have to respond to changes in their habitat to survive.

References

Ahmed, M. A. A. E. D. and Fikry, M. A. (2019). Impact of glass facades on internal environment of buildings in hot arid zone. *Alexandria Engineering Journal, 58*(3), 1063–1075.

Animalia. (2018). *The Red Junglefowl.* Retrieved from http://animalia.bio/red-junglefowl.

Anon. (1902). As long ago as the year 1661. *The Field, 99*(2558), 2.

Aonui Architecture. (n.d). *Wadestown Active Solar Home.* Retrieved from https://www.aonui.co.nz/portfolio/residential/wadestown-active-solar-home/.

Aouni Architecture. (2016). *Wadestown Active Solar Home.* Retrieved from https://www.aonui.co.nz/portfolio/residential/wadestown-active-solar-home/.

Barber, D. A. (2016). The modern solar house. pp. 1–34. *In: A House in the Sun.* Oxford Scholarship on-line.

BBC Wildlife. (2020). A year in the life of European Brown Bears. *discoverwildlife.* Retrieved from https://www.discoverwildlife.com/photo-galleries/a-year-in-the-life-of-european-brown-bears/.

Beltran, R. S., Burns, J. M. and Breed, G. A. (2018). Convergence of biannual moulting strategies across birds and mammals. *Proceedings of the Royal Society B: Biological Sciences, 285*(1878), 20180318.

Bokhorst, S., Berg, M. P., Edvinsen, G. K., Ellers, J., Heitman, A., Jaakola, L., Mæhre, H. K., Phoenix, G. K., Tømmervik, H. and Bjerke, J. W. (2018). Impact of multiple ecological stressors on a sub-arctic ecosystem: no interaction between extreme winter warming events, nitrogen addition and grazing. *Frontiers in Plant Science, 9,* 1787.

Bouâouda, H., Achâaban, M. R., Ouassat, M., Oukassou, M., Piro, M., Challet, E., El Allali, K. and Pévet, P. (2014). Daily regulation of body temperature rhythm in the camel (Camelus dromedarius) exposed to experimental desert conditions. *Physiological Reports, 2*(9), e12151.

Butti, K. and Perlin, J. (1980). *A Golden Thread.* London: Marion Boyars.

Carleton, M. A. (1914). Adaptation of the tamarisk for dry lands. *Science, 39*(1010), 692–694.

Chance, T. (2007). BedZed. Retrieved from https://www.flickr.com/photos/tomchance/1008213420.

DeKay, M. and Brown, G. (2013). *Sun, Wind, and Light: Architectural Design Strategies*. John Wiley & Sons.

Engels, S., Fong, L. S. R. Z., Chen, Q., Leng, M. J., McGowan, S., Idris, M., Rose, N. L., Ruslan, M. S., Taylor, D. and Yang, H. (2018). Historical atmospheric pollution trends in Southeast Asia inferred from lake sediment records. *Environmental Pollution, 235*, 907–917.

Everitt, B. L. (1980). Ecology of saltcedar—a plea for research. *Environmental Geology, 3*(2), 77–84.

Foruzanmehr, A. (2016). Thermal comfort and practicality: separate winter and summer rooms in Iranian traditional houses. *Architectural Science Review, 59*(1), 1–11.

Ghadirian, T., Akbari, H., Besmeli, M., Ghoddousi, A., Hamidi, A. K. and Dehkordi, M. E. (2016). Sand cat in Iran—present status, distribution and conservation challenges. *Cat News S, 10*, 56–59.

Gollin, D., Jedwab, R. and Vollrath, D. (2015). Urbanisation with and without Industrialisation. *Journal of Economic Growth, 21*(1), 35–70.

Hausladen, G. (2005). *Climate Design: Solutions for Buildings that can do More with Less Technology*. Birkhauser.

Hyde, R. (2000). *Climate Responsive Design: A Study of Buildings in Moderate and Hot Humid Climates*. Taylor & Francis.

Khajehzadeh, I., Vale, B. and Yavari, F. (2016). A comparison of the traditional use of court houses in two cities. *International Journal of Sustainable Built Environment, 5*(2), 470–483.

Lawal, R. A., Al-Atiyat, R. M., Aljumaah, R. S., Silva, P., Mwacharo, J. M. and Hanotte, O. (2018). Whole-genome resequencing of red junglefowl and indigenous village chicken reveal new insights on the genome dynamics of the species. *Frontiers in Genetics, 9*, 264.

Marsh, R. (2017). On the modern history of passive solar architecture: exploring the paradox of Nordic environmental design. *The Journal of Architecture, 22*(2), 225–251.

McLellan, B. N., Proctor, M. F., Huber, D. and Michel, S. (2017). *Ursus arctos* (amended version of 2017 assessment). The IUCN Red List of Threatened Species 2017. Retrieved from https://www.iucnredlist.org/species/41688/121229971 Accessed 1 September 2020.

Memarian, G. and Brown, F. E. (2003). Climate, culture, and religion: aspects of the traditional courtyard house in Iran. *Journal of Architectural and Planning Research*, 181–198.

Olgyay, V. (2015). *Design with Climate: Bioclimatic Approach to Architectural Regionalism-New and Expanded Edition*. Princeton University Press.

Olsson, E. A. M. (2011). *Peripheral heterothermia in reindeer (Rangifer tarandus tarandus)*. Universitetet i Tromsø,

Packer, J. E. (1964). *The Insulae of Imperial Ostia*. Berkeley: University of California.

Parr, L. (2020). The History of the Fan. Retrieved from http://www.victoriana.com/Fans/historyofthefan.html.

Phleps, H. (1982). *The Craft of Log Building: A Handbook of Craftsmanship in Wood*. Lee Valley Tools.

Polkinghorne, R. K. and Polkinghorne, M. I. R. (1949). *Other People's Houses*. London: George Harrap & Co. Ltd.

Reeder, L. (2016). *Net Zero Energy Buildings: Case Studies and Lessons Learned*. Routledge.

Richter, M. (2014). Climate aspects of the tropics. pp. 1–6. *In*: Pancel, L. and Köhl, M. (eds.). *Tropical Forestry Handbook*. Berlin; Heidelberg: Springer Nature Living Reference.

Roaf, S. (2011). *Wind Catcher*. doi:https://doi.org/10.1093/gao/9781884446054.article.T091810.

Schwartz, C. C., Miller, S. D. and Haroldson, M. A. (2003). Grizzly bear. *Wild Mammals of North America: Biology, Management, and Conservation, 2*.

Shumaker, R. W., Walkup, K. R. and Beck, B. B. (2011). *Animal Tool Behavior: The Use and Manufacture of Tools by Animals.* JHU Press.

Soppela, P., Nieminen, M. and Timisjärvi, J. (1986). Thermoregulation in reindeer. *Rangifer,* 273–278.

Stazi, F. and Naspi, F. (2018). *Impact of Occupants' Behaviour on Zero-Energy Buildings.* Springer.

Stettenheim, P. (2006). What feathers do: how many functions of feathers can you think of? The list is long and flying is just for starters. *Birder's World, 20*(3), 24.

Storey, G. R. (2002). Regionaries-type insulae 2: architectural/residential units at Rome. *American Journal of Archaeology,* 411–434.

The Fan Museum. (2020). *History of Fans.* Retrieved from https://www.thefanmuseum.org.uk/fan-history.

Vale, B. and Vale, R. (1975). *The Autonomous House.* London: Thames and Hudson.

Vale, B. and Vale, R. (1991). *Green Architecture: Design for a Sustainable Future.* London: Thames and Hudson.

Weather and Climate. (n.d.). Climate in Yazd, Iran. Retrieved from https://weather-and-climate.com/average-monthly-Rainfall-Temperature-Sunshine, Yazd, Iran.

Wheeling Jesuit University. (2004). Arctic Tundra: Plants. Retrieved from http://www.cotf.edu/ete/modules/msese/earthsysflr/tundraP.html.

Yeang, K. (2002). *Reinventing the Skyscraper: A Vertical Theory of Urban Design.* New York; Chichester: Wiley.

Chapter 3
Biomimicry and Its Approaches to Energy-Efficient Building Design

3.1 Architecture and nature: an unending dialogue

The review of examples of traditional buildings for certain climates and examples of modern energy-efficient buildings in Chapter 2, reveals the continuous attention paid by builders and architects to the surrounding environment of a building. Since ancient times, the myriad ways architects have been able to harness the potential of natural environments has evolved enormously. In modern times, these have ranged from studying the opportunities the building site provides for using passive sustainable design strategies, to more advanced ways of incorporating renewable natural energy sources for cooling, heating and ventilating buildings. The overarching aim of these approaches has been to reduce the environmental impact of a building, the need for which has been discussed in detail in Chapter 1.

Translating and incorporating biological principles into architectural design has already produced energy-efficient strategies for building design. Emulating biological structures, functions and processes has benefited the whole scope of the building life cycle comprising material production and manufacturing, design and construction, operation and demolition and recycling (Section 3.1.3.3).

Design inspired by nature has been given several names, each slightly different from the other. Even though there have been some attempts towards a theorisation of 'design inspired by nature' (Gleich 2010, Jacobs 2014), there are currently no robust definitions or agreements over the types of these design approaches and what they mean. Given this, the following section briefly presents the different terms that have been used in the field and considers the contexts in which they have usually been employed.

3.1.1 Design inspired by nature: its origins and background

Design by nature, or *bio-inspired design,* can be traced back to Leonardo Da Vinci's flying machine inspired by birds (Pohl and Nachtigall 2015). Later the bio-inspired style of Art Nouveau (1890–1910) was manifested in both architecture and the visual arts (Harris 2012). Looking at these examples, it seems both were mainly focused on the formal aspects of nature, with little attention being paid to imitating the functional principles of natural organisms (Pohl and Nachtigall 2015). This issue was addressed by Nachtigall (2010) in his book *Bionik als Wissenschaft* (Bionics as Science); a book in which the technological applications of natural principles were probably taken into account for the first time.

In 1960, Jack Steele of the Wright-Patterson Air Force Base in Dayton, OH coined the term *bionics* (Hwang et al. 2015) and pre-dated other synonyms for bio-inspired design approaches. *Bionics* has been viewed as the combination of the words of biology and technology (Gleich et al. 2010). The Merriam Webster dictionary defines *bionics* as "Science concerned with the application of data about the functioning of biological systems to the solutions of engineering problems." According to Pohl and Nachtigall (2015), the term *bionics* has existed since 1950s when the echolocation of bats was studied as inspiration for the design of radar.

The first known use of the term *biomimetics* dates back to 1969 when it appeared in an article written by Otto Schmitt (Bhushan 2009). *Bionics* and *biomimetics* have been linked to the idea of creating something artificial (Gleich et al. 2010). *Biomimetics* is involved with artificial mechanisms that produce materials similar to ones that exist in nature (Reap 2009), and is thus different from *bionic* design, which consists in taking control of nature (Wahl 2006) and which seeks to resolve engineering problems using data related to biological functions (Reap 2009). The main idea behind *biomimetics* is that no model is superior to nature when it comes to developing innovative technologies with optimal functionality and performance (Hwang et al. 2015).

The term *biomimicry* is the combination of 'biology' and 'mimicry' and was first coined by Janine Benyus (1997) in her book *Biomimicry: Innovation Inspired by Nature.* Mimicry is derived from the Greek word *mimesis* which relates to the concept of '*mimos*' and refers to "the act or the ability to simulate the appearance of someone or something else" (Marshall and Lozeva 2009). The Ancient concept *mimos* literally means *actor* or something or someone representing life. However, in the context of architectural design it tends to mean copying aspects of natural organisms. Much earlier and before the term *biomimicry* appeared, Papanek (1974) argued that *bionics* is related to *cybernetics* and Vogel (1998) claimed that it had a focus on artificial intelligence. In contrast, *biomimicry* is primarily focused on aspects of built environment sustainability (Wahl 2006) and imitating nature's efficiency (Reap 2009).

The terms *biomimicry*, *bionics* and *biomimetics* have been used interchangeably in the literature. However, *biomimicry* appears to be the more common term when it comes to architectural design. It appeared in the textbook *Bio-syllabus for European Environmental Education* that was written to give a better understanding of bio-environments and their relationship to people (Agni 2002). In the later book *Biophilic Design: The Theory, Science, and Practice of Bringing Buildings to Life* (Kellert et al. 2011), the term *biomimicry* was used to introduce a type of building design that imitates the shapes and building processes of living organisms. In this book, shells and spirals are viewed as widespread design features in the built environment which borrow their shapes and geometry from vertebrates (Heerwagen et al. 2008) (Section 3.1.3.1).

Janine Benyus argues that looking at nature and imitating its existing models, systems and processes could solve design problems in a sustainable manner (Benyus 1997). Pawlyn (2011) also suggests biological organisms can be considered as embodying technologies that offer sustainable solutions. She recommends focusing on the functional aspects of *biomimicry* rather than morphological imitations of biological samples. Technological innovations and sustainability criteria are also seen as interrelated aspects of *biomimicry*. Rao (2014) stated: "biomimicry uses an ecological standard to judge the sustainability of our innovations". *Biomimicry* has been argued to serve two main purposes, these being the production of innovation and sustainability (Pedersen Zari 2012). Bar-Cohen (2005) stated that biological processes have also been acknowledged as being very much better than human innovations.

Looking at the contexts and disciplines associated with these terms, it appears that *bionics* and *biomimetics* deal more with the technological aspects of imitating nature while the term *biomimicry* is often used where the concept of sustainability is important. This, in fact, can be seen in the early writings about the possible contribution of *biomimicry* to building design, as explored in the next section.

3.1.2 Biomimicry in architecture

Ever since the term *biomimicry* was first linked to architecture, it has been suggested it could be a design philosophy linked to industrial ecology, thus creating an interdisciplinary field having roots in the concept of 'nature as a model' (Isenmann 2002). As Allenby (1999) argued "Industrial ecology is the objective, multidisciplinary study of industrial and economic systems and their linkages with fundamental natural systems, the 'science and engineering of sustainability'."

Even though *biomimicry* was initially understood as a design philosophy for promoting sustainability, only a small number of researchers seem to have discussed its philosophical foundation (Bensaude-Vincent et al. 2002, Mathews 2011, Blok 2016, Blok and Gremmen 2016, Dicks 2016). Given this, and despite the dominant role of *biomimicry* in the industrial sector, its philosophical aspects have remained underdeveloped and descriptive (Mathews 2011).

3.1.3 Biomimicry and innovative solutions for building design

The different ways of finding inspiration in nature extend across almost all aspects of building design. This might, as suggested by Gruber et al. (2017), in part be due to the high level of complexity found in both architecture and biology. The initial stage in translating biology to architecture seems to have been lodged in copying the geometrical and morphological characteristics of living organisms. In an undeveloped approach, the aesthetic nature of living things was all that architects replicated in their design (Knippers et al. 2019). The reflection of the visual attributes of natural forms can be seen early on in the design of building ornaments and even building form. An early example of the former is reflected in the carved natural foliage on the capitals of the late 13th century Chapter House in Southwell Minster, where all of plants depicted could be seen growing in the immediate vicinity of the Minster (Figure 3-1, left). The latter approach is seen in the rib vault of the 14th century nave of Exeter Cathedral in Devon (Figure 3-1, right). Although the ribs are simply the expression of the stone arches crossing the nave that form its roof, the plethora of these at Exeter looks like a rib cage.

Given that aesthetics were the focus of these early approaches to bio-inspired design, it seems the first step in the translation of biological principles to the technical aspects of architectural design can be linked to imitating the shapes of biological organisms in building elements. Later it appears that bio-inspiration has given ideas for function, and recycling processes in buildings as well as form. As a result, this chapter

Figure 3-1. (Left) naturalistic capitals, Southwell Minster chapter house (Kognos 2019); (Right) the nave of Exeter Cathedral (DeFacto 2017).

will discuss examples under the three categories of form, function and process. Reflecting on what Frei Otto described as the key components of natural and human-made construction systems, form, function and structure work in conjunction where form and structure come into being through a developmental process (Gruber 2011). This means it would be helpful to consider biological influences on structure as part of form creation. However, in biology the concept of structure and function are also inseparably linked (Konieczny et al. 2013). This is because in any living organism, the cells organelle, cells and the tissues containing cells constitute distinct structures. Accordingly, the term 'function' refers to the sum of the functions associated with each structural level (Chauvet 2005). Therefore, here, bio-inspired structures are discussed in the function section. This has been done as it seems in architecture, structural performance is more like function than form, since the building cannot function without its structure.

Even though form, function, structure and process are integrated in the biological realm, translation of these to building design has only evolved gradually. This gradual development started from imitation of only one individual component at a time to a more hybrid method of combining these where more than one component is involved in the design by an analogy process. Sections 3.1.3.1–3.1.3.3 give examples of such bio-inspired design approaches in architecture.

Given the focus of this book is on imitating biological thermoregulation strategies, the possible future of biomimetic energy-efficient buildings could involve the exact imitation of biological processes for designing much more energy efficient and hence, sustainable buildings. Currently, the literature shows the term 'living architecture' has been assigned to different types of bio-inspired building design ranging from the simple use of green roofs and walls (Hopkins and Goodwin 2011), which apply living plants to otherwise 'dead' buildings, to a more advanced and dynamic mode of integrating life-like technologies in how a building performs (Beesley et al. 2015). In the context of this book, a living building would need to imitate the intricate hierarchical thermoregulation mechanisms of living organisms if it is to be perceived as a 'living' entity. The question then is whether this is possible, and if so, how designers can access the relevant information—this forms the subject of the later chapters in this book.

3.1.3.1 Form

The use of geometry related to the shapes of living systems can be seen in the form of several famous buildings. One such widely known form found in nature is the spiral, the mathematical principles of which have been used in designing stairs and columns. An example of the former is the bell tower staircase of the Sagrada Familia in Barcelona designed by Antonio Gaudi, while examples of the latter range from the spiral surface decoration on some of the columns of Durham Cathedral, the building of which began in the start of the 12th century, to the barley-sugar columns of the Basilica of St Peter in Rome, where the column itself appears to have been twisted into shape. More recently, Le Corbusier used spiral geometry in developing the plan of the Museum of Western Art in Tokyo, opened in 1959. The form of the 2013 Ribbon Chapel in Hiroshima by Hiroshi Nakamura and NAP is made of two interlocking spirals and the 2006 Cocoon building in Zurich by Camenzind Evolution is elliptical in shape with a floor that spirals up around a central atrium (Senosiain Aguilar 2003).

A more complicated translation of biological forms into architectural design seems to have happened through the discovery by Ernst Haeckel (1834–1919) of the biogenic law, which led to the morphology of organisms being scientifically considered for the first time. In his book *General Morphology of Organisms*, Haeckel outlined an evolutionary morphology through classification of forms during the initial stages of development in living organisms (Breidbach 2006). The influence of Haeckel's theory and its reflection in architectural design can be seen in the design of an entrance for the World Exhibition Paris 1900 (Figure 3-2, left). The French architect Rene Binet was inspired by the geometry of a microscopic single-celled marine animal illustrated in Haeckel's book, *Report on the Radiolaria*

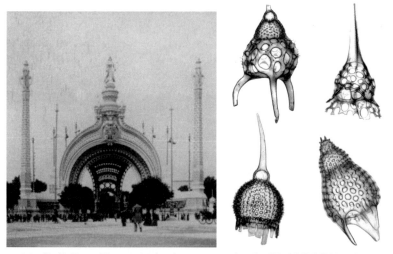

Figure 3-2. (Left) Porte Monumentale: the entrance for the World Exhibition Paris 1900 (Brown University 1900) (Source: Image provided by the authors from (https://upload. wikimedia.org/wikipedia/commons/1/13/Porte_Monumentale_de_l%27Exposition_ universelle_1900.jpg)), (Right) species of radiolarium (Author, adapted from (Picturepest 2014, 2016, 2017, Drews 2016)).

(Figure 3-2, right) (Proctor 2006). Hersey's (2001) description below of Rene Binet's design, known as *Porte Monumentale*, shows the link between the gateway's proposed structure and function, and Radiolaria (a protozoa) as the source of inspiration for the design:

> "…gateway is a tightly swollen lattice curved into a huge, three-legged truss, with two legs in front and one behind. The gate, again like the animal, forms an open, webbed dome. As a further bow to Oceania, Binet's structure is outlined with arrays of light bulbs beaming light to the world in the manner of a sea creature's organs of luminescence."

Early Art Nouveau designers were also inspired by the organic forms presented in the book *Art Forms of Nature* by Haeckel (Sabin and Jones 2017). The prominent figures of the Art Nouveau movement, Victor Horta and Hector Guimard adopted such forms in their projects. The stairway of Hotel Tassel (1882–1883) (Figure 3-3, left) and the entrance to Porte Dauphine metro station in Paris (1889) (Figure 3-3, right) are examples of such projects. Later, in USA, Sullivan's use of botanical organic ornaments was poetic and similar to the curves used in Art Nouveau (Dennis and Wenneker 1965).

Frank Lloyd Wright (1953), who shaped the organic architecture movement states "Organic means Part-to-Whole-as-Whole-is-to-Part." Organic architecture considers a building and its surrounding environment as an integrated whole similar to a living being, within which parts are integrated and work in unity (Senosiain Aguilar 2003). Organic architecture is conceived as being rooted in a passion for life, nature and natural forms. It has also been linked to the process of growth and change known as 'metamorphosis' in which architectural design starts within and from a seed which then grows outward to create forms (Pearson 2001). Such description of organic architecture appeared in the essay "In the Cause of Architecture" published by Wright in 1914. The influence of biological forms in Wright's

Figure 3-3. (Left) Main stair of Hotel Tassel in Brussels by Victor Horta (Townsend 2002); (Right) Porte Dauphine metro station (Bellomonte 2012).

projects can be seen in the design of the 1959 Guggenheim Museum in New York, where he used the spiral as the generator of the building interior in the form of a ramp that coils up towards the glass ceiling with the art displayed along it.

Digital morphogenesis or *computational morphogenesis* is a much more modern and advanced method of generating forms for buildings. It uses the principles of evolutionary development in biological organisms for generating complex building forms (Lynn 2000). A comparatively undeveloped example for this is the design of the late 1990s Embryological House in which the form-finding process is metaphorically linked to the growth and development of the embryo (McGinley 2015).

3.1.3.2 Function

The internal and external complexity of biological structures has been reflected in the design of bio-inspired building structures, while adaptation of biological organisms to their surrounding environments has informed principles for environmental building design. This means bio-inspired design strategies have been used to improve both the structural and environmental performance of a building as well as demonstrate how the building functions.

The concept of building performance was first introduced by Blachere (1965) and led to its introduction in building in 1972 as defined by the International Council for Building Research Studies and Documentation (CIB) Commission (Foster 1972). Building performance can be evaluated from the three perspectives of (1) the requirements of the occupants, (2) economics and (3) environmental performance (Leaman et al. 2010). The structural performance of a building seems to be related to the first and possibly second items while the second and third relate to the characteristics of buildings with regards to their environmental impact from the perspective of energy and material flows (Lützkendorf et al. 2005). The following sections briefly explain how the translation of biological principles has affected structural and environmental aspects of building performance.

3.1.3.2.1 Structural performance

In the design of bio-inspired building structures which imitate not only the geometry and macro or structural organisations of living systems but also the structural elements within the load-bearing network, it seems the initial inspiration has come from looking at forms and functions working in conjunction. In this area, D'Arcy Thompson's book, *On Growth and Form*, probably had a significant influence on understanding biological forms in plants and animals through mathematics, which is essential for comprehending forms and their functions. According to his research, the generation of biological forms was driven by the internal and external dynamic forces coming from either within cells and tissues or from the immediate environment (Sykes 2010).

Gaudi's approach to using organic forms was not simply based on copying but rather on mathematical properties. Being impressed by Ruskin and Viollet Le Duc, he designed structures in which shape, space and function were appropriately synchronised. His inspirations mainly came from the skeletons of animals and the biomechanical characteristics of plants (Senosiain Aguilar 2003).

Buckminster Fuller took a different approach to studying organic structures. Compared to Gaudi's focus on the roles of natural forces in shaping forms, Fuller's interest was in studying the optimised structures and load-bearing systems found in biological structures such as muscles and bones (Ingber 2003). He defined his 'tensegrity structure' as a system that stabilises its shape through continuous tension.

A combination of the principles found in the man-made constructions and structural processes in nature was manifested in Pier Luigi Nervi's projects. Nervi's roof forms display a balanced merging of the geometrical and mathematical aspects of organic and orthogonal structures (Leslie 2003).

Mies van der Rohe, a pioneer of the modern movement owned a copy *On Growth and Form* and his skeletal forms for Lakeside Drive appear to have been influenced by D'Arcy Thompson's analogies between natural and man-made skeletal structures. The reflection of this inspiration can be seen in the work of his student Maryon Goldsmith, whose thesis *The Tall Building: The Effects of Scale*, explored the relationship between scale and efficiency in nature and the ways it can be translated into structural design (Leslie 2003).

A later example of form-finding processes under the influence of extrinsic forces in nature can be seen in Frei Otto's designs where the self-organisation of material systems was used as a technique for creating load-bearing structural systems (Menges 2006). Antoni Gaudi and Frei Otto's designs were similar in terms of using gravity as the fourth dimensional non-Euclidean parameter in their flexible physical models that were used to explore architectural form (Figure 3-4, left).

"Although Buckminster Fuller and Frei Otto significantly advanced both the theory and propaganda promoting pneumatic structures, others applied and commercialized these principles" (LeCuyer 2008). An example is the series of temporary snack bars with pneumatic roofs designed by Victor Lundy for the 1964/65 New York World's Fair (Figure 3-4, right).

Gaudi's design method (Figure 3-5) has been described as the conventional way of achieving parametric design (Burry 2016). The deep understanding of natural forces and the integration of material and structure at both the macro and micro levels by Gaudi and Frei Otto were

Figure 3-4. (Left) Otto's study of bubbles (Author, adapted from LeCuyer (2008)); (Right) One of Victor Lundy's pneumatic roofs, New York World's Fair (Author, adapted from LeCuyer (2008)).

Figure 3-5. Tree-like column structure in Sagrada Familia.

reflected in their designs for structural systems. The hyperbolic paraboloid, which is an infinite ruled surface discovered in the 17th century (Chen and Peng 2013), was used by Gaudi in several projects.

The use of computers and structural analysis software has facilitated structural modelling, load analysis, and thus, the design and construction of complex and massive structures. A modest example of using Computer Aided Design (CAD) can be seen in the design of the spiral form of the Stammheim Church in Germany by Peter Hubner, who was known to be influenced by the organic architecture movement. His design, however, used the mathematical idea of the logarithmic spiral to develop timber structural forms using uniform B-spline curve modelling using C "...with each timber element being different in terms of span, angle, and height" (Szalapaj 2014).

A more advanced method of designing biomorphic shell structures led to the roof of the Centre Pompidou in Metz designed by Shigeru Ban (Figure 3-6, left, right). The shell structure is more complicated as it involved non-uniform B-spline curve modelling. Construction of the timber segments was achieved through digital fabrication using the generative methods within the computer software technology. Even though the form of the roof is organic and functional, the structural elements of the shell do not seem to have been directly inspired by a living organism. The grid shell structure expands on Frei Otto's designs for massive grid shells.

Calxton Fidler (1841–1917) made a comparison between animal skeletons and the main girder of a cantilever bridge (Figure 3-7). Much more recently many of the structures of the Spanish architect/engineer Santiago Calatrava resemble the skeletons of birds and animal. The imitation of the gross morphology and structure of skeletal systems can also lead to building structural elements that have an almost dynamic appearance, examples of which are Calatrava's Milwaukee Art Museum in Milwaukee, Wisconsin (1994–2001) and the World Trade Center Transportation Hub in New York City (2016), with its resemblance to a bird about to take flight.

Technological improvements have provided opportunities for scientists to study the structure, function, and motion of the mechanical aspects of biological systems at different levels, from the whole body of

Figure 3-6. The Center Pompidou in Metz. (Left) (Lucie2beaugency 2012); (Right) (Dalbéra 2010).

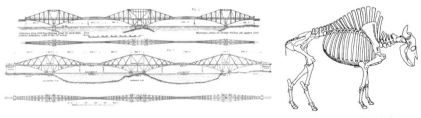

Figure   3-7. Similarities   between   man-made   structures   and   animal   skeletons. (Left) (Westhofen 1890); (Right) (Grunwald 2016, adapted from Thompson 2014).

the organism, to organs, cells, and tissues. Research organisations such as PlanktonTech in Germany are recognised for their exclusive research on the biomechanical characteristics of natural structures (Pohl 2015). They use computational design to replicate the optimisation capabilities of light-weight biological structures to develop new products for building structures and facades.

As discussed in Section 3.1.3.1, the preliminary approach to morphogenesis in architecture mainly focused on form generation, irrespective of the structural functionality of that form. A more technical approach to the concept of morphogenesis in architecture was discussed in relation to research conducted by Roudavski (2009). In this, digital computation was used as a means of employing the rules of developmental morphology and growth dynamics in plants to create functional structures for buildings. In morphogenetic architecture, generation of the final form is achieved through a procedural transformation in which iterations of forms are derived from the extrinsic forces.

3.1.3.2.2 Environmental performance: operation

One of the approaches to improving the environmental performance of a building is to reduce its energy use as buildings consume 40% of total global energy consumption (Sustainable Buildings and Climate Initiative (SBCI) 2009). Climatic conditions and building physics affect building performance (Chen 2015) as disassociation between indoor and outdoor conditions leads to higher energy consumption so as to keep indoor conditions within the comfort range.

People adapt to the climates in which they live, so that people in Alaska might be in shirt sleeves in temperatures that would see someone living in Hong Kong wearing a padded jacket and woolly hat. The problem with many modern buildings is that they take no account of outdoor conditions and set the indoor environment to a certain temperature bandwidth, such as that set by ASHRAE, that is supposed to fall within the comfort zone. Comfort studies have also confirmed this tendency to be true (Humphreys 1978, Turner 2011), and revealed that setting inappropriate indoor temperatures and thus dissociating indoor and outdoor conditions can lead to higher than necessary energy consumption (Chen 2015). This means adaptation of a building's indoor environment to the surrounding environment could reduce energy use and make a significant contribution to climate change mitigation. The key here is that animals and plants have adapted to their environments and hence it seems reasonable to examine how they do this in terms of thermoregulation.

Recent research shows that this type of biomimicry promises to improve environmental conditions in buildings (Pedersen Zari 2010, Pawlyn 2011, Badarnah 2012, Gamage and Hyde 2012, Mazzoleni 2013). There is also evidence that imitating the different ways biological organisms use to adapt to their environment has already led to reduced energy use in a building. The examples below show how biomimetic design has led to the reduction of energy use both for the construction of a building and its operation.

The design of the Eastgate Centre in Harare, Zimbabwe was inspired by termite mounds (Figure 3-8, left). Despite being a popular example of an energy-efficient biomimetic building, it has received criticism (Turner and Soar 2008). The suggested anomaly in the design that separates it from how a termite mound works comes from the need for huge fans to move air which consume a significant amount of energy to run. Natural movement of air in a termite mound occurs through the temperature difference between the nest and the outside air (see Chapter 4 for details) (Figure 3-8, right).

Biological thermal adaptation strategies have been classified into three groups (Dregne 1999, Sejian et al. 2012). These are morphological, behavioural, and physiological and translating the principles of any of these could suggest energy-efficient approaches to building design.

Looking at the Eastgate building, it seems that it has imitated the morphological aspects of a termite mound, but that the spaces through which the air is directed in the building are very much simpler than the different sizes of spaces in a termite mound (Figure 3-8, right). In the building, the space is dictated by the building function, in other words, the need to house office workers. Given this, it appears the physiological and behavioural aspects of living organisms have not yet been fully investigated and transferred to architectural practice, and the purpose of

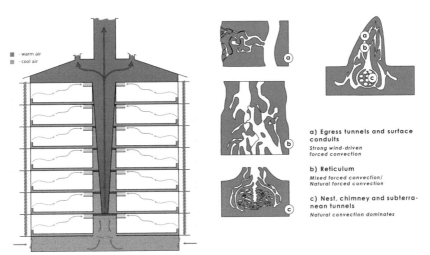

Figure 3-8. (Left) Natural fan-assisted ventilation in Eastgate Centre (Author, adapted from Oyster (2010)); (Right) Functional organization of a termite mound (Author, adapted from Turner and Soar (2008)).

this book is to find a way of accessing these and translating them so they can be useful in the latter.

3.1.3.3 *Process*

Earlier, a mention was made of the fact that bio-inspiration has been applied to processes like recycling in buildings. There is no concept of waste in nature but human systems have not as yet been able to imitate such closed cycles on any scale. Thus, the design and construction process may also be able to draw on inspiration from the natural world in this regard. Chapter 2 described how traditional building processes are often formed by the restricted availability of local materials. Here, the focus is on the modern design and construction processes, particularly in relation to the choice of materials and what happens to these at the end of a building's life.

3.1.3.3.1 Environmental performance: material production and construction processes

The life cycle of a building has to consider all stages from the choice of materials and how these are extracted and/or made and transported to the site, to the materials required for maintenance and the energy to run the building in the use phase, and finally, what happens when the building is no longer needed and its materials enter the waste stream. All of these stages have an impact on the environment. The embodied energy in a building comprises the initial energy in its materials and that in materials used for its maintenance and possible refurbishment, with the former being recognised as the more significant contributor to carbon emissions (Treloar et al. 2001, Rauf and Crawford 2015). Initial embodied energy (IEE) is the sum of the energy used for mining and processing the natural resources, manufacturing the building materials, transportation of raw resources to factories and manufacturing sites and product delivery to the building site and the energy used for the construction activities. The energy used for subsequent refurbishment, repair, replacement and maintenance of components or materials during the service life of a building is known as recurrent embodied energy (REE).

The literature shows that the use of bio-inspired materials has contributed to reducing initial embodied energy in buildings. An example is materials based on natural honeycomb structures. The unique properties of the honeycomb is strength and lightness, so less material can do the same job. A simple example is the standard flush door, where two thin skins of plywood are separated by a honeycomb paper structure that provides stiffness having much less weight, and hence, requiring less material, than an equivalent solid timber door.

While people have always been fascinated by honeycomb structures, only in the last 70 years have they found large-scale applications in the form of papers, metals, ceramics and composites (Zhang et al. 2015). The current materials most used for honeycomb fabrication are aluminium, polymers and composites (Vitale et al. 2017), such as found in sandwich panels made to improve thermal performance. However, looking to biology to improve building materials, beyond the use of so-called natural materials like straw and timber, is an area yet to be fully explored and one beyond the remit of this book with its focus on operating energy efficiency.

3.1.3.3.2 Demolition process

The fourth stage of a building's life cycle is the demolition stage. Demolition is the process of dismantling the structure and dealing with the unwanted components. This can be a process of separating them as salvage in the hope they can find new uses, although most demolition material is treated as construction waste (Udawatta et al. 2015, Park and Tucker 2017, Bakchan and Faust 2019).

Research shows that about 50% of the embodied energy can be recovered through recycling (Thormark 2001). Like construction energy, the energy used for demolition makes up a small share of the total energy a building consumes during its lifetime, the majority of which for non-low energy buildings comes from the operating energy. Taking operating energy into account, the maximum energy recovered through recycling would be up to 15% of a building's total energy use (Thormark 2002).

The Cradle to Cradle (C2C) approach was an effort to reduce waste by going beyond the concept of eco-efficiency to eco-effectiveness (McDonough and Braungart 2002). While, in the former, the aim is to reduce the environmental impact of products, the latter is a shift towards eliminating waste, just as happens in the natural world. The C2C design framework consists of a biological cycle and a technical cycle, both being a means of reducing waste. The biological cycle was inspired by the zero-waste characteristics of nature and thus uses the flow of biological materials in ecosystems as a model for creating a similar flow for industrial materials. The technical cycle, on the other hand, concerns the everlasting durability of industrial materials once they are produced. For example refined copper and aluminium, although they have an environmental impact during their production, can be recycled indefinitely, although some energy and new material may have to be added to maintain product quality.

Modern construction has embraced the use of bio-based materials as a means of achieving sustainability (Bardage 2017). 'Bio-based materials' indicate the origin of a material, such as timber or bamboo since these are grown, while 'recyclable', 'biodegradable' and 'compostable' are terms that describe what happens at the end of the life of a material. This suggests not all bio-based materials are recyclable. Timber is a useful example as it can be both reused and recycled but ultimately, it is also compostable as it will rot.

Bio-inspired materials are synthetic materials, the properties of which emulate the biological principles of living matter. The use of biological principles in the design of bio-inspired materials can be seen at the macro, micro and nano levels. Bio-based materials, often referred to as natural materials, are different from bio-inspired materials. The reason is that bio-based materials by definition consist of at least one substance derived from living matter with either vegetal, animal, bacterial or fungal origins (biomass). The living substance (biomass) could be used unmodified, as in the case of reed used for thatching, modified as in the conversion of a log to building timber, or as bio-composites which are made from combining one or more materials (Jones 2017). The latter might be as simple as a mud brick reinforced with chopped straw. All materials in these categories are classified as being bio-based. A sub category of bio-based materials, are bio-aggregate-based in which plants are added to a material matrix, such as hemp concrete (Amziane and Collet 2017).

While the concept of imitating thermoregulation in nature seems unrelated to bio-based materials, there is a relationship as all bio-based materials contain living matter. This means that the valued performance characteristics of such materials which are derived through their evolutionary processes can be exploited and used in the building industry (Jones 2017). For example, timber in a building can lock up carbon for the life of the building, thus leading to a reduction in GHG emissions. If the timber components can be reused in another building then the carbon is locked up for longer. Bio-based recyclable materials occur as bioplastics and bio-composites and a brief explanation of these is provided below.

Bioplastics may be polymers where the feedstock for their manufacture is from a renewable source, such as algae, and some will also be biodegradable, but this depends on their chemical structure rather than the type of feedstock, so not all bioplastics are biodegradable (Mekonnen et al. 2013). Hydrocarbons in the form of proteins, fats (from animals) and cellulose, oils and starches (from plants) can be used to make bioplastics. Stabnikov et al. (2015) state, "The use of biodegradable plastics in construction reduces the land for disposal of construction wastes after demolition and diminishes the cost of construction works because the biodegradable plastic foams, sheets, liners, and fences can be left in soil without their excavation and disposal."

Bio-composites are manufactured materials that contain natural fibres (bio-fibres). Wood-polymer composites have been used for skirting board profiles and even for scaffolding (Grigorieva and Oleinik 2016). Cork particles have been combined with polypropylene at an experimental level to change the physical properties of the latter (Fernandes et al. 2014), although cork already has a role in the building industry as a flooring material and insulation. Another experiment used cork sand and cement to create a non-load-bearing insulating panel (Boussetoua et al. 2017). Jute-based composites show high tensile strength and are widely used in ceilings, floors and windows (Asha et al. 2017) while hemp concrete has both good thermal insulating properties and good moisture transfer and storage abilities (Aït Oumeziane et al. 2017).

References

Agni, V.-A. (ed.). (2002). *Bio-syllabus for European Environmental Education: A Textbook for the Better Understanding and Appreciation of the Bio-environment.* Biopolitics International Organisation.

Aït Oumeziane, Y., Moissette, S., Bart, M., Collet, F., Pretot, S. and Lanos, C. (2017). Influence of hysteresis on the transient hygrothermal response of a hemp concrete wall. *Journal of Building Performance Simulation, 10*(3), 256–271. doi:10.1080/19401493.2016.1216166.

Allenby, B. R. (1999). *Industrial Ecology and Design for Environment.* Paper presented at the Proceedings First International Symposium on Environmentally Conscious Design and Inverse Manufacturing.

Amziane, S. and Collet, F. (2017). *Bio-aggregates Based Building Materials: State-of-the-Art Report of the RILEM Technical Committee 236-BBM.* Dordrecht: Dordrecht: Springer Netherlands.

Asha, A. B., Sharif, A. and Hoque, M. E. (2017). Interface interaction of jute fiber reinforced PLA biocomposites for potential applications. pp. 285–307. *In: Green Biocomposites.* Springer International Publishing.

Badarnah, L. (2012). *Towards the Living Envelope: Biomimetics for Building Envelope Adaptation.* (Doctoral dissertation), TU Delft, Zutphen, Netherlands.

Bakchan, A. and Faust, K. M. (2019). Construction waste generation estimates of institutional building projects: Leveraging waste hauling tickets. *Waste Management, 87*, 301–312.

Bar-Cohen, Y. (2005). *Biomimetics: Biologically Inspired Technologies.* CRC Press.

Bardage, S. (2017). Performance of buildings. pp. 335–383. *In: Performance of Bio-based Building Materials.* Elsevier.

Beesley, P., Chan, M., Gorbet, R., Kulić, D. and Memarian, M. (2015). Evolving systems within immersive architectural environments: new research by the living architecture systems group. *Next Generation Building, 2*, 31–56.

Bellomonte. (2012). Paris Metro 2 Porte Dauphine Libellule. Retrieved from https://commons.wikimedia.org/wiki/File:Paris_Metro_2_Porte_Dauphine_Libellule. JPG, Creative Commons CC0 1.0 Universal Public Domain Dedication: https://creativecommons.org/publicdomain/zero/1.0/deed.en.

Bensaude-Vincent, B., Arribart, H., Bouligand, Y. and Sanchez, C. (2002). Chemists and the school of nature. *New Journal of Chemistry, 26*(1), 1–5. doi:10.1039/b108504m.

Benyus, J. (1997). *Biomimicry: Innovation Inspired by Nature.* New York, NY: Morrow.

Bhushan, B. (2009). Biomimetics. *In: The Royal Society.* London.

Blachere, G. (1965). *General Consideration of Standards, Agreement and the Assessment of Fitness for Use.* Paper presented at the The 3rd CIB Congress on Towards Industrialised Building, Copenhagen, Denmark.

Blok, V. (2016). Biomimicry and the materiality of ecological technology and innovation: toward a natural model of nature. *Environmental Philosophy, 13*(3), 195–214.

Blok, V. and Gremmen, B. (2016). Ecological innovation: biomimicry as a new way of thinking and acting ecologically. *Journal of Agricultural and Environmental Ethics, 29*(2), 203–217. doi:10.1007/s10806-015-9596-1.

Boussetoua, H., Maalouf, C., Lachi, M., Belhamri, A. and Moussa, T. (2017). Mechanical and hygrothermal characterisation of cork concrete composite: experimental and modelling study. *European Journal of Environmental and Civil Engineering*, 1–16. doi:10.1080/19648 189.2017.1397551.

Breidbach, O. (2006). The conceptual framework of evolutionary morphology in the studies of Ernst Haeckel and Fritz Müller. *Theory in Biosciences, 124*(3-4), 265–280.

Brown University. (1900). La porte monumentale, Exposition Universelle 1900. Retrieved from https://commons.wikimedia.org/wiki/File:La_porte_monumentale,_Exposition_ Universelle_1900.jpg, public domain: https://en.wikipedia.org/wiki/Public_domain, {{PD-US}} – U.S. work that is in the public domain in the U.S. for an unspecified reason, but presumably because it was published in the U.S. before 1926.

Burry, M. (2016). Antoni Gaudi and Frei Otto: Essential precursors to the parametricism manifesto. *Architectural Design, 86*(2), 30–35.

Chauvet, G. (2005). *The Mathematical Nature of the Living World: The Power of Integration.* World Scientific.

Chen, J. and Peng, F. (2013). Approximate spline of G2-continuity on a generalized hyperbolic paraboloid. *Journal of Computational and Applied Mathematics, 248*, 99–117.

Chen, Y. (2015). *Environmental Adaptive Design Building Performance Analysis Considering Change.* (Master's), University of South Carolina, Colombia, SC (10799532).

Dalbéra, J.-P. (2010). Le centre Pompidou Metz. Retrieved from https://www.flickr.com/ photos/dalbera/4913713833/in/photolist-Gvgp3-5hoLC9-5xe3nC-dSJ13C-8JrVvL- eWTT8y-8AB3PZ-9zBWQY-cWuLeL-6gEzud-78QXH8-7VyC8J-oPEBji-8ud4W2- aYpTGP-pDESDp-6zETbh-7ofS2N-r6JYE4-8J9xKo-pmWga4-iFGDBE-6AER1t-9Ta8S7- 7Vvkft-5V5YCG-rAMofj-8udWWR-6weqfp-pENDmS-hwh5kj-6wiEYs-8vs2pb- nHkDgr-qiERiq-azgPGT-pDniQ3-78QXsk-d1tEvN-rmwgYJ-pnvepC-atPkLT-6wetoX- 5V1v4t-d1ee4A-5HAWkx-8BoyKg-8vgLfw-6w5X7v-d1hwfy, Creative Commons: https://creativecommons.org/licenses/by/2.0/.

DeFacto. (2017). Exeter Cathedral nave vaulted ceiling. Retrieved from https://commons. wikimedia.org/wiki/File:Exeter_Cathedral_nave_vaulted_ceiling.jpg, Creative Commons Attribution-Share Alike 4.0 International license: https://creativecommons. org/licenses/by-sa/4.0/deed.en.

Dennis, J. M. and Wenneker, L. B. (1965). Ornamentation and the organic architecture of Frank Lloyd Wright. *Art Journal, 25*(1), 2–14. doi:10.1080/00043249.1965.10793756.

Dicks, H. (2016). The philosophy of biomimicry. *Philosophy & Technology, 29*(3), 223–243. doi:10.1007/s13347-015-0210-2.

Dregne, H. (1999). Pan. *Encyclopedia of Deserts. University of Oklahoma Press. Norman*, 499–500.

Drews, A. (2016). Radiolarian - Podocyrtis (Lampterium) mitra Ehrenberg - 160x. Retrieved from https://commons.wikimedia.org/wiki/File:Radiolarian_-_Podocyrtis_ (Lampterium)_mitra_Ehrenberg_-_160x.jpg, Creative Commons Attribution 2.0 Generic license, https://creativecommons.org/licenses/by/2.0/deed.en.

Fernandes, E. M., Correlo, V. M., Mano, J. F. and Reis, R. L. (2014). Polypropylene-based cork–polymer composites: processing parameters and properties. *Composites Part B: Engineering, 66*, 210–223.

Foster, B. E. (ed.). (1972). *Performance Concept in Buildings* (Vol. 361). Philadelphia, PA: National Bureau of Standards 361.

Gamage, A. and Hyde, R. (2012). A model based on Biomimicry to enhance ecologically sustainable design. *Architectural Science Review, 55*(3), 224–235. doi:10.1080/00038628. 2012.709406.

Gleich, A. (2010). *Potentials and Trends in Biomimetics* (1st ed. 2010. ed.). Berlin, Heidelberg: Springer Berlin Heidelberg.

Gleich, A., Pade, C., Petschow, U. and Pissarskoi, E. (2010). *Potentials and Trends in Biomimetics*. Springer Science & Business Media.

Grigorieva, L. and Oleinik, P. (2016). *Recycling Waste Wood of Construction*. Paper presented at the Materials Science Forum.

Gruber, P. (2011). *Biomimetics in Architecture: Architecture of Life and Buildings*. Wien: Springer.

Gruber, P., McGinley, T. and Muehlbauer, M. (2017). Towards an agile biodigital architecture: Supporting a dynamic evolutionary and developmental view of architecture. *In: Interdisciplinary Expansions in Engineering and Design With the Power of Biomimicry*. IntechOpen.

Grunwald, A. M. (2016). Bison Skeleton. Retrieved from https://commons.wikimedia.org/wiki/File:Bison_Skeleton.svg, Creative Commons Attribution-Share Alike 4.0 International license: https://creativecommons.org/licenses/by-sa/4.0/deed.en.

Harris, S. (2012). *Nikolaus Pevsner: The Life*. London, UK: Random House.

Heerwagen, J., Kellert, S. R. and Mador, M. (2008). *Biophilic Design: The Theory, Science, and Practice of Bringing Buildings to Life*. Hoboken, N.J: Wiley.

Hersey, G. L. (2001). *The Monumental Impulse: Architecture's Biological Roots*. Mit Press.

Hopkins, G. and Goodwin, C. (2011). *Living Architecture: Green Roofs and Walls*. Csiro Publishing.

Humphreys, M. (1978). Outdoor temperatures and comfort indoors. *Batiment International, Building Research and Practice, 6*(2), 92–92.

Hwang, J., Jeong, Y., Park, J. M., Lee, K. H., Hong, J. W. and Choi, J. (2015). Biomimetics: forecasting the future of science, engineering, and medicine. *International Journal of Nanomedicine, 10*, 5701.

Ingber, D. E. (2003). Tensegrity I. Cell structure and hierarchical systems biology. *Journal of Cell Science, 116*(7), 1157–1173.

Isenmann, R. (2002). Further efforts to clarify industrial ecology's hidden philosophy of nature. *Journal of Industrial Ecology, 6*(3-4), 27–48.

Jacobs, S. (2014). Biomimetics: a simple foundation will lead to new insight about process. *International Journal of Design & Nature and Ecodynamics, 9*(2), 83–94.

Jones, D. (2017). Introduction to the performance of bio-based building materials. pp. 1–19. *In: Performance of Bio-based Building Materials*. Elsevier.

Kellert, S. R., Heerwagen, J. and Mador, M. (2011). *Biophilic Design: The Theory, Science and Practice of Bringing Buildings to Life*. John Wiley & Sons.

Knippers, J., Schmid, U. and Speck, T. (eds.). (2019). *Biomimetics for Architecture: Learning from Nature*. Basel: Birkhäuser.

Kognos. (2019). Southwell Minster Carvings Chapter House Tympana Retrieved from https://commons.wikimedia.org/wiki/File:Southwell_Minster_Carvings_Chapter_House_Tympana_07-11_Capitals_09-14.jpg, Creative Commons Attribution-Share Alike 4.0 International license: https://creativecommons.org/licenses/by-sa/4.0/deed.en.

Konieczny, L., Roterman-Konieczna, I. and Spólnik, P. (2013). *Systems Biology: Functional Strategies of Living Organisms*. Springer Science & Business Media.

Leaman, A., Stevenson, F. and Bordass, B. (2010). Building evaluation: practice and principles. *Building Research & Information, 38*(5), 564–577.

LeCuyer, A. (2008). *ETFE: Technology and Design*. Walter de Gruyter.

Leslie, T. (2003). Form as diagram of forces: the equiangular spiral in the work of Pier Luigi Nervi. *Journal of Architectural Education, 57*(2), 45–54.

Lucie2beaugency. (2012). Metz Musée G Pompidou intérieur Poutres de bois. Retrieved from https://commons.wikimedia.org/wiki/File:Metz_Mus%C3%A9e_G_Pompidou_

int%C3%A9rieur_Poutres_de_bois.jpg, Creative Commons Attribution-Share Alike 4.0 International license: https://creativecommons.org/licenses/by-sa/4.0/deed.en.

Lützkendorf, T., Speer, T., Szigeti, F., Davis, G., le Roux, P. C., Kato, A. and Tsunekawa, K. (2005). A comparison of international classifications for performance requirements and building performance categories used in evaluation methods. *In*: P. Huovila (ed.). *Performance Based Building*. Espoo, Finland: Technical Research Centre of Finland.

Lynn, G. (2000). Embryologic houses. *Architectural Design, 70*(3), 26–35.

Marshall, A. and Lozeva, S. (2009). Questioning the theory and practice of biomimicry. *International Journal of Design & Nature and Ecodynamics, 4*(1), 1–10.

Mathews, F. (2011). Towards a deeper philosophy of biomimicry. *Organization & Environment, 24*(4), 364–387. doi:10.1177/1086026611425689.

Mazzoleni, I. (2013). *Architecture Follows Nature: Biomimetic Principles for Innovative Design*. Boca Raton, FL: CRC Press.

McDonough, W. and Braungart, M. (2002). *Cradle to Cradle: Remaking the Way we make Things* (1st ed.). New York, NY: North Point Press.

McGinley, T. (2015). A morphogenetic architecture for intelligent buildings. *Intelligent Buildings International, 7*(1), 4–15.

Mekonnen, T., Mussone, P., Khalil, H. and Bressler, D. (2013). Progress in bio-based plastics and plasticizing modifications. *Journal of Materials Chemistry, 1*(43).

Menges, A. (2006). Polymorphism. *Architectural Design* (180), 78–87.

Nachtigall, W. (2010). *Bionik als Wissenschaft: Erkennen-Abstrahieren-Umsetzen*. Heidelberg, Germany: Springer.

Oyster, F. t. (2010). Natural ventilation high-rise buildings. Retrieved from https://commons.wikimedia.org/wiki/File:Natural_ventilation_high-rise_buildings.svg, Creative Commons Attribution-Share Alike 3.0 Unported: https://creativecommons.org/licenses/by-sa/3.0/deed.en.

Papanek, V. (1974). *Design for the Real World: Human Ecology and Social Change*. London, UK: Thames and Hudson.

Park, J. and Tucker, R. (2017). Overcoming barriers to the reuse of construction waste material in Australia: a review of the literature. *International Journal of Construction Management, 17*(3), 228–237.

Pawlyn, M. (2011). *Biomimicry in Architecture*. London, UK: RIBA Publishing.

Pearson, D. (2001). *New Organic Architecture: The Breaking Wave*. Univ of California Press.

Pedersen Zari, M. (2010). Biomimetic design for climate change adaptation and mitigation. *Architectural Science Review, 53*(2), 172–183. doi:10.3763/asre.2008.0065.

Pedersen Zari, M. (2012). *Ecosystem Services Analysis for the Design of Regenerative Urban Built Environments*. (Doctor of Philosophy), Victoria University of Wellington, Wellington, New Zealand.

Picturepest. (2014). Radiolaria (fossile) - 400x (negative photo) (14175878555). Retrieved from https://commons.wikimedia.org/wiki/File:Radiolaria_(fossile)_-_400x_(negative_photo)_(14175878555).jpg, Creative Commons Attribution 2.0 Generic license, https://creativecommons.org/licenses/by/2.0/deed.en.

Picturepest. (2016). Thyrsocyrtis sp. - Radiolarian (30442423343). Retrieved from https://commons.wikimedia.org/wiki/File:Thyrsocyrtis_sp._-_Radiolarian_(30442423343).jpg. Creative Commons Attribution 2.0 Generic license, https://creativecommons.org/licenses/by/2.0/deed.en.

Picturepest. (2017). Anthocyrtium hispidum Haeckel - Radiolarian (34986365113). Retrieved from https://commons.wikimedia.org/wiki/File:Anthocyrtium_hispidum_Haeckel_-_Radiolarian_(34986365113).jpg, Creative Commons Attribution 2.0 Generic license, https://creativecommons.org/licenses/by/2.0/deed.en.

Pohl, G. (2015). *Biomimetics for Architecture & Design Nature-Analogies-Technology* (1st ed. 2015. ed.). Cham: Springer International Publishing.

Pohl, G. and Nachtigall, W. (2015). *Biomimetics for Architecture & Design: Nature-Analogies-Technology*. Springer.

Proctor, R. (2006). Architecture from the cell-soul: René Binet and Ernst Haeckel. *The Journal of Architecture, 11*(4), 407–424.

Rao, R. (2014). Biomimicry in architecture. *International Journal of Advanced research in Civil, Structural, Environmental and Infrastructure Engineering and Developing, 1*(3), 101–107.

Rauf, A. and Crawford, R. H. (2015). Building service life and its effect on the life cycle embodied energy of buildings. *Energy, 79*, 140–148.

Reap, J. (2009). *Holistic Biomimicry: A Biologically Inspired Approach to Environmentally Benign Engineering*. (Ph.D.), Georgia Institute of Technology, Atlanta, GA.

Roudavski, S. (2009). Towards morphogenesis in architecture. *International Journal of Architectural Computing, 7*(3), 345–374.

Sabin, J. E. and Jones, P. L. (2017). *Labstudio: Design Research Between Architecture and Biology*. Routledge.

Sejian, V., Naqvi, S., Ezeji, T., Lakritz, J. and Lal, R. (2012). *Environmental Stress and Amelioration in Livestock Production*. Springer.

Senosiain Aguilar, J. (2003). *Bio-architecture*. Oxford: Architectural.

Stabnikov, V., Ivanov, V. and Chu, J. (2015). Construction biotechnology: a new area of biotechnological research and applications. *World Journal of Microbiology and Biotechnology, 31*(9), 1303–1314.

Sustainable Buildings and Climate Initiative (SBCI), U. (2009). Buildings and climate change: Summary for decision-makers. Retrieved 8 May 2018 https://europa.eu/capacity4dev/unep/document/buildings-and-climate-change-summary-decision-makers.

Sykes, A. K. (ed.). (2010). *Constructing a New Agenda Architectural Theory 1993–2009*. New York: Princeton Architectural Press.

Szalapaj, P. (2014). *Contemporary Architecture and the Digital Design Process*. Routledge.

Thompson, D. W. (2014). *On Growth and Form*. Cambridge University Press.

Thormark, C. (2001). Conservation of energy and natural resources by recycling building waste. *Resources, Conservation and Recycling, 33*(2), 113–130.

Thormark, C. (2002). A low energy building in a life cycle—its embodied energy, energy need for operation and recycling potential. *Building and Environment, 37*(4), 429–435.

Townsend, H. (2002). Tassel House stairway. Retrieved from https://commons.wikimedia.org/wiki/File:Tassel_House_stairway.JPG, Public Domain: https://en.wikipedia.org/wiki/Public_domain.

Treloar, G. J., Love, P. E. and Holt, G. D. (2001). Using national input/output data for embodied energy analysis of individual residential buildings. *Construction Management and Economics, 19*(1), 49–61.

Turner, J. S. and Soar, R. C. (2008). *Beyond Biomimicry: What Termites can tell us About Realizing the Living Building*. Paper presented at the First International Conference on Industrialized, Intelligent Construction, Loughborough, UK.

Turner, S. C. (2011). What's new in ASHRAE's standard on comfort. *ASHRAE Journal, 53*(6), 42.

Udawatta, N., Zuo, J., Chiveralls, K. and Zillante, G. (2015). Improving waste management in construction projects: An Australian study. *Resources, Conservation and Recycling, 101*, 73–83.

Vitale, J. P., Francucci, G. and Stocchi, A. (2017). Thermal conductivity of sandwich panels made with synthetic and vegetable fiber vacuum-infused honeycomb cores. *Journal of Sandwich Structures & Materials, 19*(1), 66–82.

Vogel, S. (1998). *Cats' Paws and Catapults: Mechanical Worlds of Nature and People* (1st ed.). New York, NY: Norton.

Wahl, D. C. (2006). Bionics vs. Biomimicry: From control of nature to sustainable participation in nature. pp. 289–298. *In*: C. A. Brebbia (ed.). *Comparing Design in Nature with Science and Engineering* (Vol. 87). Southampton, UK: WIT.

Westhofen, W. (1890). Forth Bridge (1890). Retrieved from https://commons.wikimedia. org/wiki/File:Forth_Bridge_(1890)_Fig._4,_Page_5.png, Public Domain: https:// en.wikipedia.org/wiki/Public_domain.

Wright, F. L. (1953). *The Language of Organic Architecture*. Taliesin square-paper 16: Spring Green, Wisconsin.

Zhang, Q., Yang, X., Li, P., Huang, G., Feng, S., Shen, C., Han, B., Zhang, X., Jin, F., Xu, F. and Lu, T. J. (2015). Bioinspired engineering of honeycomb structure–Using nature to inspire human innovation. *Progress in Materials Science, 74*, 332–400.

Chapter 4
Linking Biology and Buildings

4.1 The search for a link between biomimetic design and building energy efficiency

As a first step, a literature review was conducted to explore whether and how biomimetic design has already been linked to reducing building operational energy. The aim was to discover how researchers have been inspired by biological thermal adaptation mechanisms and the innovative architectural thermoregulation strategies these have produced, and also to identify any methods for finding appropriate thermoregulatory solutions in nature.

Four databases were used at the start of the search and Table 4-1 sets out the terms used and the number of results obtained. References of the most promising articles were also screened.

Table 4-1. Search terms within the databases Google Scholar, Sage, Scopus and Science Direct: (date of the last search: 28 July 2016).

Database	Keywords	Number of Results
Google Scholar	('thermoregulation' AND 'Biomimicry' AND 'architecture')	2350
	(thermal regulation AND Biomimicry AND architecture) NOT ('medical' [All Fields]) NOT ('morphogenesis' [All Fields]) NOT ('material' [All Fields])	177
Sage	All Fields ('thermal regulation' AND 'biomimicry')	11
Scopus	All Fields ('thermal regulation' AND 'biomimicry' AND 'architecture')	63
	TITLE-ABS-KEY ('thermal/thermo regulation' AND 'biomimicry')	1
Science Direct	Field of Engineering ('thermal regulation' AND 'biomimicry' AND 'architecture')	14

The three keywords: "medical", "morphogenesis" and "material" were excluded in the searches as they identified papers that were not relevant. Morphogenesis in architecture is the study of biological forms with the aim of developing environmentally responsive forms and structures. The mathematical rules governing biological structures have in turn inspired the design and development of construction materials and structural elements, and the generation of forms as described in Chapter 3. Although bio-inspired materials could be developed with the aim of reducing the embodied energy of buildings, morphogenesis was excluded from the keywords as the focus of this investigation was building operational energy.

This extensive list of publications needed further investigation in terms of their relevancy. Usefulness was determined by reading abstracts and sometimes whole articles. Eventually 203 papers were selected for further investigation (177 from Google Scholar, 11 from Sage, 1 from Scopus and 14 from Science Direct). Screening these, three duplicates were found. Eventually, 127 papers were dropped based on their titles and a further 44 were dropped after reading the abstracts because these did not exactly address the aim of this research. Irrelevant papers dealt with the topics of tissue engineering (medical science), cellular biology, the role of biomimicry in the future of cities (not relevant because of the scale of the research), using the principles of the organisation of ecological communities (ecosystems) for creating sustainable human communities, papers focused on computation and genetic algorithms, bio-inspired structures for buildings and bio-inspired textiles. This left 29 papers for detailed study.

The selected articles dealt with the following key concepts: thermo/ thermal regulation in buildings, biomimetic, biomimicry, bio-inspired architecture, bio-inspired design, natural organisms, biological samples, thermal adaptation strategies/mechanisms, energy efficiency, thermal performance, heat balance strategies in nature and heat regulation mechanisms of plants/humans/animals. The articles were then analysed for inclusion in a comparative table (Table 4-2) based on the following steps:

Step 1: Is the article a one-off study with no description of how the biomimetic analogy was found, and no use of the same analogy in other design situations (Y = Yes, N = No)?

Step 2: "Practical" means the energy efficiency of the case study has been analysed using energy simulation; "theoretical" implies this step is omitted (T = Theoretical, P = Practical).

Step 3: Extraction of the name of the software used to evaluate the usefulness of the biological thermal adaptation strategies on the thermal performance of the redesigned/bio-inspired buildings.

Table 4-2. Summary of research linking biomimicry and thermal performance.

	Step 1	Step 2	Step 3	Step 4	Step 5	Step 6	Step 7	Step 8
(Park and Dave 2014)	Y	P	Ecotect	Animal eyes	Daylight distribution	Adaptable façade systems	NA	N
(Wang and Li 2010)	Y	T	NA	Butterfly	Adaptive response to solar radiation	Parametric kinetic building envelope	NA	N
(Wang 2011)	Y	T	NA	Butterfly	Climatic temperature responsiveness	Parametric kinetic building envelope	CC	N
(Zare and Falahat 2013)	Y	T	NA	Reptiles	Introduces several thermoregulation strategies beneficial for architectural design	Conceptual smart wall	HC	N
(Tachouali and Taleb 2014)	Y	P	Ecotect	Flamingo	Indoor space temperature	Suggests strategies for reducing total energy consumption	H&AC	N
(Zuazua-Ros et al. 2016)	Y	P	Open Studio & Energy Plus	Tuna	Reducing heating demand of the building	Heat management in building	DC	N
(Al Amin and Taleb 2016)	Y	P	Ecotect+ ladybug	Desert snail	Enhancing thermal comfort	Reducing energy consumption	HC	N
(Bermejo-Busto et al. 2016)	Y	T	NA	Beehives	Better control of the air cavity	Redesigning Peltier cell prototypes	NA	N
(Ahmar and Fioravanti 2014a)	Y	P	Energy Plus + DIVA	Termites	Daylight control	Suggests strategies for reducing building energy consumption in a hot climate	H&AC	N

Table 4-2 Contd. ...

...Table 4-2 Contd.

	Step 1	Step 2	Step 3	Step 4	Step 5	Step 6	Step 7	Step 8
(Worall 2011)	Y	T	NA	Termites	Energy efficiency and ventilation	Distribution of HVAC systems	NA	N
(Turner and Soar 2008)	Y	T	NA	Termites	Airflow regulation/ adjusting natural ventilation	Suggests a theoretical model for distribution & control of HVAC systems	NA	N
(Badarnah and Knaack 2007)	Y	T	NA	Sea sponge	Regulating the interior microclimate	Conceptual breathing skin	DC	N
(Han et al. 2015)	Y	P	Open Studio & Energy Plus	Flower petals	Reducing urban heat island effect	Retro-reflective building envelope	H&AC	N
(Alkhateeb and Taleb 2015)	Y	P	IES	Tree bark	Thermal insulation	Suggests strategies for reducing total energy consumption	HC	N
(Nanaa and Taleb 2015)	Y	P	Ecotect	Lotus flower	Use of natural sustainable energy	Suggests strategies for reducing total energy consumption	NA	N
(Lee 2008)	Y	T	NA	Hair	Mitigating heat load	Overcoming thermal issues	NA	N
(Reddi et al. 2012)	Y	T	NA	Plant leaves & Skin	Stabilized earth construction	NA	NA	N
(Lopez et al. 2015)	Y	T	NA	Plant stomata	Adaptive envelope	Suggests active materials for the building envelope	DC	N
(Lopez et al. 2015)	Y	T	NA	Plant stomata	Adaptive envelope	Suggests active materials for the building envelope	NA	N

Reference				Biological source	Investigation	Application		
(Reichert et al. 2015)	Y	T	NA	Plant cones	Investigates the use of hygroscopic material for the building facade	Weather responsive composite	DC	N
(Alston 2015)	Y	T	NA	Chemical composition of tree surface	Investigates the possibility of developing a material with adaptive real-time performance	Intelligent glass surfaces	NA	N
(Nessim 2015)	Y	P	ANSYS	Human skin	Airflow performance with adaptive response to temperature without the use of mechanical or electrical means	Responsive material	H&AC	N
(Scartezzini et al. 2015)	N	T	NA	Animals	Effective control over indoor climatic conditions	Suggests strategies for reducing total energy consumption	H&AC	N
(Badarnah 2012)	N	T	NA	Organisms	Evaporative cooling system in building envelopes	Design of a building facade	H&AC	N
(Ahmar and Fioravanti 2014a)	N	P	DIVA	Plants	Improving the thermoregulation of the building skin	Parametric shading screen	H&AC	N
(Ahmar and Fioravanti 2014b)	N	T	NA	Plants	Cooling in hot climates	Categorization of biomimetic ideas in thermal regulation based on a small database	H&AC	N
(Elghawaby 2010)	N	T	NA	Flora & fauna	Improving the ventilation system	Conceptual breathing facades	H&AC	N

Table 4-2 Contd. ....

...Table 4-2 Contd.

	Step 1	Step 2	Step 3	Step 4	Step 5	Step 6	Step 7	Step 8
(Kim and Torres 2015)	N	P	Design builder/ EnergyPlus	Perforated folded surface of organisms	Exploring an integrated sustainable façade system	Contemporary curtain wall glazing system	NA	N
(Badarnah and Kadri 2015)	N	T	NA	Random organisms	Creating a biophysical framework for accessing relevant analogies	Provides part of a structured framework of heat regulation processes	H&AC	Y

Step 4: Finding the names of the natural organism(s) used as design inspiration under the three categories of animals, plants and human anatomy.

Step 5: Extracting the architectural objectives mentioned.

Step 6: Identifying the architectural solution(s) or technologies used to overcome thermal issues.

Step 7: Labelling the climate context of the research (NA = Not Available, CC = Cold climate, HC = Hot climate, H&AC = Hot and arid climate, DC = Different climates).

Step 8: Indication of whether there were explicit links to previous data collections or strategies for identifying relevant thermal adaptation strategies found in nature (Y = Yes, N = NO).

4.2 Extraction of useful data

The 29 articles were then categorised into three types: one-off examples, varied examples and flora and fauna. The first was the dominant type with 22 of the 29 articles dealing with one-off examples. This suggested it might be more useful to look at the type of biological inspiration, whether from animals and insects, plants or human beings. The results are discussed below.

4.2.1 *Animals and insects*

Park and Dave (2014) studied the geometrical and functional characteristics of the compound eyes of shrimps, which are made up of repeating units acting as visual receptors. Their aim was to design the surface of a gymnasium for badminton using a series of adjustable reflectors so that the amount of light entering could be controlled. This responsive facade system was designed to be capable of adapting itself to the environment by following the sun path. Modelling revealed daylighting was improved and the energy consumption dropped, although the proposed system required expensive automatic sensors potentially leading to high-maintenance costs. The emulation process was also limited to mimicking the formal configuration of a natural organism.

The bio-inspired kinetic envelope (BKE) suggested by Wang and Li (2010) was designed to change its form in response to exterior change. Thus, the envelope would open up and concentrate the solar radiation and transfer it to the interior but become smooth at night and in winter to minimise the heat lost from the surface. The authors claimed the wings of butterflies were the inspiration for this idea. The micro-structures on butterfly wings can both receive and block solar radiation as necessary. However, this research was an idea for a façade shape and there was no thermal analysis to support the claim for optimisation of energy consumption. The translation of biological principles to architectural parameters here still follows a morphological approach.

Of the studies examined in this group, 17 did not provide enough evidence to prove that biomimicry could effectively enhance thermal regulation in buildings. However, the biomimetic approach of Zare and Falahat (2013) is useful. They found that reptiles adapt to their environment using one or multiples of the following methods: water evaporation from the skin, hibernation in winter and aestivation in summer, an excretion system that retains water, solar energy absorption, increasing the length of their hands and feet and sheltering in the ground. They consequently suggested a number of architectural solutions such as underground construction and reducing a building's surface to volume ratio to improve the control of ventilation, heat absorption and humidity with the aim of reducing energy consumption in a hot and arid climate.

In this animal and insect group, 11 studies had run energy simulations to test the energy efficiency of the biomimetic design approaches. Tachouali and Taleb (2014), using a school in Dubai as a case study, translated the breathing mechanisms of the flamingo into "a flow of fresh air within the corridors and [to] create a cross ventilation in the classrooms", although cross ventilation is hardly a new idea. The bird's feathers also inspired the design of louvres for shading, and the flamingo's food filtering system led to the design of a buffer to channel and filter wind. The designs were simulated using Ecotect and although there were reductions in energy use, the authors stated "mixed-mode ventilation is still needed to achieve optimal thermal comfort."

Zuazua-Ros et al. (2016) used OpenStudio and showed that the heating demand of a workplace could drop significantly if space were to be arranged based on the strategy used by tuna to conserve energy. This was tested for three different climates by manipulating the two variables of different types of floor layout and occupancy level. It emerged that the location of meeting rooms or individual offices in the inner area of the office building (analogous to the location of dark muscles in the body of a tuna) had a considerable effect on the energy savings (Figure 4-1).

Al Amin and Taleb (2016) were inspired by the desert snail and some of its thermal survival mechanisms were applied to a box-like three-storey research building in Dubai. The building and the improvements were modelled using Ecotect. The desert snail has a white shell to reflect solar radiation; changing the reflectivity and emissivity of the roof of the building reduced the cooling load by 4%. When it is hot, the snail retreats to the upper part of the shell leaving a layer of air under it to insulate it from the hot ground. Thus, improving the building's insulation reduced the cooling load by 19%. The building was also modelled with a new glass

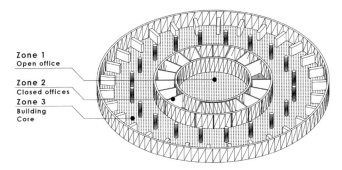

Zone 1
Open office

Zone 2
Closed offices

Zone 3
Building
Core

Figure 4-1. The tuna option for the floor plan of offices (Author, adapted from Zuazua-Ros et al. (2016)).

wall separating a circulation zone adjacent to the outside wall from the sedentary office space, rather in the way the snail retreats within its shell. This reduced the cooling load by 3%. Overall, the most effective method was insulating the building, which although inspired by the desert snail is not a new architectural idea.

Bermejo-Busto et al. (2016) proposed a new HVAC system based on the Peltier effect but following biomimetic principles. The Peltier or thermoelectric effect is based on two dissimilar materials which, when an electric current is passed through them, develop a hot side and a cool side. If the hot side is attached to a heat sink, the heat will be dissipated and the cool side can drop to below room temperature, meaning the effect can be used for cooling interior spaces. In this HVAC system, heat is extracted from the interior and passed to cavities in the façade in the hot season. This was inspired by the heat shield created between the brood comb and the exterior wall of the beehive (Figure 4-2, left). Heat-shielding is a temperature regulation behaviour used by bees to insulate the brood from localised heat stress. Siegel et al. (2005) conducted a series of tests to determine the contribution and behaviour of stationary and moving workers on keeping the brood cool. In response to heat stress, they found the number of worker bees increases on both the hive wall and the brood, and the movement of those on the hive wall increases to improve the airflow (Figure 4-2, right).

Of the studies reviewed, a number were related to imitating the thermal regulation strategies found in termite mounds. For example, Ahmar and Fioravanti (2015) suggested the use of a porous double skin façade in the Egyptian climate, thus imitating a feature of the termite mounds found in the Sahara Desert. The porous nature of the surface means that wind driven air can enter the mound and circulate through the conduits in the surface whatever the wind direction. The porous surface layers also protect the inner nest from turbulent winds. The study followed a morphological

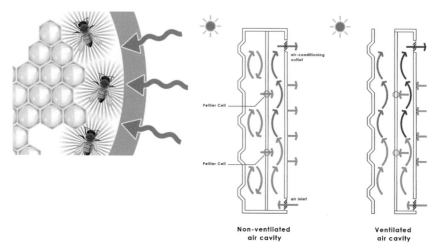

Figure 4-2. Thermal behaviour of bees. (Left) heat shield (Author, adapted from Pixabay (2017) and Freepic (n.d.)); (Right) Peltier System and cavity performance in winter and summer (Author, adapted from Bermejo-Busto et al. (2016)).

approach by manipulating the geometry of the façade. This consisted of a hierarchy of three iterations of a triangular pattern, and by altering a series of attributes such as the fold depth and area of the triangles, the degree of cooling could be changed. The design was simulated using EnergyPlus v.8.2 for the thermal and DIVA for the daylight simulations. One room in an existing building with and without the new double porous façade design was modelled. With the façade, there was a 15% reduction in cooling load but also a 45.2% reduction in daylighting, suggesting further design modifications would be needed.

Worall (2011) discussed the working of the termite mound in more general terms and suggested mimicking the vascularized and tidal ventilation of termite mounds could lead to improved thermal performance in buildings. He also warned that the "…termite mound has many other functions apart from providing ventilation, and so picking one out may result in inappropriate applications", going on to state that it was better to understand the principles behind the functioning of a natural organism like a termite mound and then apply these principles to building.

The Eastgate Centre and its relationship to the thermoregulation of termite mounds has been described in Chapter 3. Turner and Soar (2008) noted that termite mounds do not require forced-air plants to regulate the temperature. Further, they stated there were similarities between termite mounds and lungs, and a good understanding of termite mounds and human lung functions could help architects design more sustainable buildings without the need for high-energy consuming HVAC systems.

4.2.2 *Plants*

Plants have also inspired new ideas for thermoregulation in buildings. The accumulation of buildings in urban environments leads to the urban heat island effect, which in turn increases building energy consumption through more need for cooling (Oke 1973, Imhoff et al. 2010, Loughner et al. 2012). Differing from researchers who have analysed individual building energy performance rather than that of groups of buildings in an urban context, Han et al. (2015) looked for a biomimetic solution for reducing the impact of urban heat islands. They found the flower *Galanthus Nivalis* (common snowdrop) that appears at the end of winter in countries like the UK, produces a cooler intrafloral temperature through the retro-reflective surface of its petals. Based on this, a folded facade was designed for a building located in an urban context in eight USA cities representing six major climate zones (Figure 4-3). The retro-reflective façades led to building energy reductions in all cities but the authors also suggested "…that retro-reflective building envelopes might have more impact in warmer climatic cities that have longer daylight time and more demand for cooling energy."

Alkhateeb and Taleb (2015) studied the complex structure of tree bark to find design strategies for a residential villa situated in Dubai. The strategies they extracted were translated into design principles such as shading, cavity walls and thermal insulation. Using these, energy consumption in the optimal bio-inspired model was reduced by 34%. A similar approach was taken by Nanaa and Taleb (2015) who investigated the strategies used by the lotus to cope with its tropical environment. Both studies suggested double glazed windows and the use of a shading system as bio-inspired design strategies, but again these are not new strategies for reducing energy use in buildings.

A number of research studies have taken a different approach towards creating a plant-inspired adaptive building envelope by investigating the

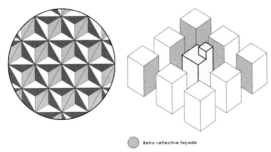

Retro-reflective façade

Figure 4-3. (Left) The pattern used for designing retro-reflective building façades (Author, adapted from Han et al. (2015)); (Right) Building blocks (Author, adapted from Han et al. (2015)).

direct use of natural materials. These are again theoretical studies that have not involved testing the energy efficiency of the 'plant-inspired building'. López et al. (2015) stated that biomimetic buildings inspired by stomatal movement would be able to behave like a living organism, that is, by responding to the environment without requiring any sensors or actuators. In land-based plants, stomata are the pores that control the flow of water and CO_2 in and out of the leaf. Highly occupied spaces in buildings, such as lecture rooms, can lead to a build-up of CO_2 and water vapour, which is normally controlled by artificial or natural ventilation. The aim was to find materials that could act more like stomata in leaves. López et al. (2015) began by investigating the changes in CO_2 and water within a leaf when environmental parameters such as humidity, temperature, CO_2 concentration and light intensity changed. Finally, they suggested a number of reactive materials for windows and tested these in the laboratory by monitoring the dimensional changes in the window surfaces. These materials were heat-sensitive plastics, wood and carbon dioxide responsive polymers. However, the materials were not tested in real buildings and the potential energy savings were not simulated. The same team also applied the adaptive behaviour of stomata in terms of thermal regulation, water management and air exchange to create adaptive building envelopes, while accounting for different climate zones and building user demands. Ultimately, they suggested the use of hydrogels and thermal polymers as these could produce real-time responses in the envelope to environmental stimuli (López et al. 2015). However, the suggestions remain to be tested in the field.

Research conducted by Alston (2015) investigated the hierarchical structure in the material composition of tree bark, as the tree surface is capable of regulating solar adsorption. The absorbency and adaptive real-time performance of tree bark was examined in order to design an imitative intelligent glass façade with a vascular or fluid circulation network similar to that of the bark. This was intended to act as an adaptive cooling layer for the building envelope. The chemical fluid would flow through the vascular networks in the glass, which had a similar pattern to that of tree bark. Experiments were set up and sensors were installed in the experimental glass façades to record the temperature fluctuations. The flow of chemical fluid in the glass was manipulated to enable the façade to either absorb or obstruct solar energy as required. Alston added that this bio-inspired glass façade would reduce the urban heat island effect. This work remains to be tested at the large scale.

Reichert et al. (2015) suggested another responsive architectural system which did not use electronic equipment for sensing, actuation and control. Their examination of the literature revealed that most current approaches to making a building responsive to climatic issues were based on electrical and mechanical systems. In contrast, they believed that imitating the hygroscopic actuation plants employ for organ movement would allow for developing a passive, material-embedded responsiveness. This led to the design of two responsive materials composed of composite cells called Hygroscope and Hygroskin. Each cell has two timber layers of which one opens and the other closes in response to changes in humidity. This highly adaptive building envelope is limited to one specific type of material (wood) and one environmental variable (moisture). The proposed system has yet to be used in a real building.

4.2.3 *Human beings*

Human skin has always been a source of inspiration for thermal regulation in buildings. Gruber and Gosztonyi (2010) aligned the functions of the skin of organisms with analogies in architecture: "Skins, shells, cuticles, membranes and other outer layers of organisms protect, confine and contain living organisms. Architecture delivers an internal environment for mostly human activity, and facades create a difference between the inside and the outside." Other researchers have considered buildings envelopes as analogous to human skin in terms of controlling heat generation and maintaining thermal equilibrium (Badarnah and Knaack 2007, Reddy et al. 2007, Lee 2008, Nessim 2015).

Lee (2008) believed building envelopes have similarities with human skin in terms of controlling heat generation and maintaining thermal equilibrium. He describes the function of the hair follicle and how it facilitates heat regulation by either lying flat on the skin or being erect in cold conditions to trap a layer of still air as insulation between the skin and external air. He developed various designs for building envelopes but the energy efficiency of their application to buildings was not evaluated.

Badarnah and Knaack (2007) proposed a new type of exterior wall capable of inhaling and exhaling, and thus adapting to the changing environment. The proposed façade consisted of elements using an elastic membrane that can deform to allow air to flow in or out as required. The exterior wall is made up of these repetitive components or chambers that are similar to human lungs. The air pressure inside a chamber is controlled by wires and valves attached to its expelling surface (Figure 4-4, left). This mechanism regulates the airflow through the building. The efficiency of such a breathing wall remains unknown, however, as no simulation has been run to evaluate its performance. Moreover, not enough information is provided on the technical implications of this conceptual design.

Elghawaby (2010) proposed a breathing wall for buildings in the Sinai Peninsula (Figure 4-4, right). This used the ability of the human skin to sweat as a cooling mechanism as inspiration for the design. The exterior layer allows the air to pass through while also preventing direct sunlight

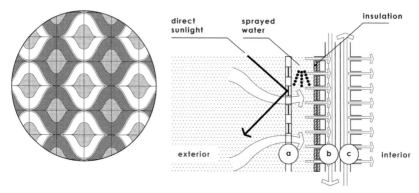

Figure 4-4. (Left) Repetitive components of the façade that are similar to human lungs (Author, adapted from Badarnah and Knaack (2007)); (Right) "Breathing wall" (Author, adapted from Elghawaby (2010)).

from entering the building. The middle layer works as an insulation layer in which air is cooled by a water spray system outside the skin. The internal layer has ventilation outlets for bringing the cooled air inside and these can be controlled as necessary. This is a very rare case where a bio-architectural solution could be generally applicable to all buildings in similar climates and that also has the potential to be further technologically developed. However, the energy efficiency of the proposed system remains unknown.

Reddi et al. (2012) looked at heat generation and evaporative heat loss as the thermoregulatory mechanisms of human skin and plant leaves and suggested these represented a series of challenges and opportunities for biomimetic construction that might be exploited in the design of stabilised soil walls. They stated, "The ideas of embedding plant fibers, not only just for reinforcement but also for thermoregulatory purposes, and imitating biological systems such as vascular network in skin, remain to be explored." As an analogy, the capillaries in the soil through which water is distributed were claimed to be similar to sweat glands in skin. Similarly, the shrink-swell nature of soil was linked to the vasoconstriction/vasodilation of the blood vessels. Most people building reinforced earth walls want to stop water passing from outside to inside and having walls that swell when it rains might not be useful. However, Reddi et al. (2012) claimed walls of stabilised earth could be engineered to imitate biological principles and hence, reduce energy consumption. However, they also acknowledged that for the claim to become reality, the disciplines of biology, soil science, construction and engineering would need to come together.

Only one paper in this category proved the energy efficiency of the proposal using simulation, that of Nessim (2015). Starting from the view that the building envelope is analogous to human skin, this study analysed the thermal performance of a shape memory polymer (SMP) using an experimental, physical model of two conjoined rooms with an internal window between them. The air flow through the model building was improved by the 2 mm sheets of SMP used in its internal window. These sheets bent outward to allow for airflow when the temperature inside the model was increased, thus allowing more ventilation from the air coming through the window in the external wall.

4.2.4 Other relevant examples

A number of studies have discussed the idea of bio-inspired thermal regulation in buildings by looking at multiple biological examples. Scartezzini et al. (2015) studied the thermoregulation strategies of silkworms, honey bees and tuna and foresaw a number of possible bio-inspired projects ranging from passive design strategies to manipulation of control networks. Based on an interpretation of a building as a cold-blooded animal, a number of passive functions were suggested for providing a comfortable situation for the inhabitants, such as a heat recovery system for Spanish office buildings based on the thermal behaviour of tuna, but these ideas were neither well explained nor developed. Badarnah et al. (2010) studied human skin, tuna fish and termite mounds by looking for the features, challenges, strategies and mechanisms these organisms use for thermal adaptation. This led to a proposal for a new form of brick—the Stoma Brick. This was a façade that used the principle of evaporative cooling, although it was designed to operate in hot humid and hot dry climates, with a cycle appropriate for cold weather. This is a design yet to be tested. Sara and Noureddine (2015) developed a bio-key tool, which is based on biomimetic principles inspired by prairie dogs, spiders and termite mounds. The tool set out to improve the energy consumption of a building by reducing the energy used for cooling, ventilation and improving the indoor air quality. However, there is no evidence of whether energy use can be optimised emulating these principles. The proposed bio-inspired design framework that forms their tool is also not an original methodology as it is based on the BioGen system developed by Badarnah (2015), and requires pre-existing knowledge of natural organisms.

Ahmar and Fioravanti (2014b, 2014a) pointed out how the thermoregulation strategies of botanical examples (tree bark, leaves and succulents) can be analysed and categorized into four groups each representing a specific type of heat transfer method that could be used as inspiration for new architectural features. They also explained how the size, shape, orientation and ventilation system of the surface of the leaf affects thermoregulation. They suggested the building façade could be made of panels each oriented slightly differently with regard to the incident solar radiation. They suggested that, "On a micro scale, the surface properties of these panels could have a rough texture to diffuse incident light and provide self shading." Their later design for a double skin façade based on thermoregulation in termite mounds (Ahmar and Fioravanti 2015) has been discussed earlier in this chapter.

An integrated façade system for a multi-storey office building in Charlotte, North Carolina, was designed to reduce heat loss in cold seasons and increase it in hot seasons by optimising the daylight entering through it throughout the year (Kim and Torres 2015). The design was loosely based on the ideas behind biomimicry. The authors listed examples of biological processes that provided inspiration for the design of a façade for a sustainable building as "…photosynthesis, the hydrophobic effect, and photocatalysis, which were reflected and developed as a photovoltaic façade, self-cleaning façade, and pollutant removal façade, respectively." The folded perforated skin consisted of aluminium components which followed the movement of the sun. Based on the simulation results, the total energy consumption of the baseline building façade was 30% more than that of the integrated façade.

4.3 A systematic way of accessing natural examples of thermoregulation

Table 4-3 summarises the examples outlined in this chapter and their relationship to architecture. These examples did not appear to have emerged as the result of a systematic search for possible strategies found in nature that might have architectural applications. However, it seemed there were examples of how imitating thermal adaptation processes in nature had the potential to contribute to sustainable building design, and that designers could translate the biological strategies they found into architectural design principles.

Given the underlying concept of this book is the importance of having a framework that allows innovative ideas to be found and translated into design, it seems that most researchers do not follow a well-developed bio-inspired design (BID) process when it comes to finding relevant solutions in nature for overcoming thermal challenges buildings. The BID approach has been given multiple names such as 'problem-based', 'challenge to biology' and 'technology pull', all referring to a bio-inspired design approach that begins with a design challenge and then explores nature to find solutions. In the context of energy-efficient biomimetic building design inspired by thermoregulation processes in nature, only Badarnah (2012) has developed a framework for the BID process. Her BioGen framework, which grew out of her PhD research, is described and discussed below.

Table 4-3. Summary of examples.

Biological example	Nature	Strategy	Architecture	Strategy
Termite mounds	Capped chimneys	Air refreshing process Water evaporation Respiratory gases exchange	A row of tall stacks	Cooling mechanism
	Open-chimney mounds	Stack effect mechanism	Flow of heat	Cooling mechanism
	Damping temperature & fan-driven ventilation	Thermal capacity of the soil	Low/high capacity fans	Cooling mechanism
Tree bark	Protection	Tree bark double layer	Multilayer skin building envelope	Cooling/heating mechanism
	Self-shading	Peeling crusts, deep cracks	Shading	Cooling mechanism
	Insulation	A waxy thick layer of cork prevents water loss in plants	Thermal insulation materials	Cooling/heating system
	Material conductivity	Inner bark responsible for transporting sugars	High-performance glazed units	Cooling/heating mechanism
Lotus flower	Light control	Mechanism of opening and closing depending on daylight level	Shading louvres	Cooling mechanism
	Self-shading	The higher leaves provide shading for the flower	Double roof	Cooling mechanism
	Nourishment	Water and CO_2 in the presence of light are converted to glucose	PV cells	Protection from heat gain
Flower petals	Evaporation/ self-shading	Special directional reflective property of flower petals	Retro-reflective building envelope	Cooling mechanism
Desert snail	Withdrawal inside the shell		Cool roof/ thermo-shield	Insulation
	Shell reflectivity		Reflective surfaces	Reduction in heat gain

Table 4-3 Contd. ...

...Table 4-3 Contd.

Biological example	Nature	Strategy	Architecture	Strategy
	Shade	Low temperature of the soil surface under the snail	Intermediate zones	Insulation
	Air as insulation	Withdrawal of snail inside the shell	Extending the roof	Shading
	Evaporative cooling	The weight of a snail changes from day to night	Night purging of building atrium	Cooling mechanism
Fish	Preventing glare	Polarization of light	Glass material	Light distribution
Moth	Glare reduction	Anti-reflective due to a protrusion on the surface of the cornea	Glass material	Light distribution
Lobster	Glare reduction	Square tubes	Glass material	Light distribution
Black stork	Glare reduction	The relative angle between water surface, and beak	Glass material	Light distribution
Fly	Glare reduction	Absorbing light by red colouring	Glass material	Light intensity
Meerkat	Glare reduction	Absorption of light by black fur surrounding their eyes	Glass material	Light intensity

4.3.1 BioGen (a biomimetic framework for design concept generation)

In her thesis, Badarnah (2012) addresses the need for a framework through which architects could generate design concepts. This is based on her belief that drawing on the innovative adaptation mechanisms employed by natural organisms will result in building envelopes that are capable of adapting themselves to the environment by regulating the four main environmental aspects of heat, water, light, and wind. The purpose of her research was thus to provide a method for generating architectural design concepts. Doing this would enable designers to go beyond merely imitating the morphology of organisms and move biomimetic architectural design towards a more functional-oriented approach.

Her research also addressed a series of sub-themes:

1) It investigated the environmental criteria that affect the building envelope and subsequently, the indoor climatic conditions.

2) It sought relevant adaptation strategies in nature that could be used in building envelopes.

3) It explored a method for representing the strategies offered by nature.

4) It developed a method for generating design concepts based on the lessons suggested by relevant organisms.

It seems these sub-themes were developed to achieve the main objective of the research, which was how the design principles, methods and strategies identified in natural organisms could be abstracted and transformed into design concepts for regulating building envelopes that could provide good interior thermal comfort without the excessive use of fossil fuel energy. To reduce the research down to a manageable level, only a small number of relevant natural organisms were to be included in the BID framework (Badarnah 2012). As a result, the functional aspects of the thermal adaptation methods she used for establishing BioGen are somewhat limited, as compared to the huge gamut of survival strategies found in nature. In terms of assuring the generalisability of BioGen, Badarnah (2012) proposed the design concept generation framework should be tested for the four different environmental principles of temperature, light, wind and air. Through regulating these the internal environments of buildings should maintain homeostasis similar to that achieved by organisms.

The challenges encountered during her research are set out below:

1) How to search for and select adaptation strategies.

2) The applicability of the strategies in design, as some might work in only one of the three chosen scales (macro, micro and nano). The three scales refer to the levels at which thermal adaptation takes place as thermal regulation can be achieved at organismal, cellular, or molecular level.

3) The conflict between the different solutions a design concept might need in various parts.

The BioGen framework consisted of two main steps, which were referred to as the preliminary design phase and the emulation phase, although Badarnah stated the second phase was not considered in her thesis. Figure 4-5 shows the sub-phases developed for BioGen. The light grey rectangles introduce the sub-phases developed for the preliminary design and the dark grey are the sub-phases of the emulation phase. The different shades of light grey represent the three main steps of the

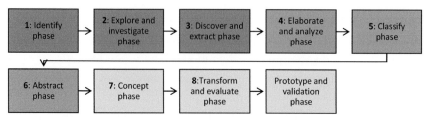

Figure 4-5. Outline of the design methodology map developed by Badarnah (Author, adapted from Badarnah (2012)).

bio-inspired building envelope framework. The darkest light grey is the exploration phase, the mid-light grey is the pinnacle analysis ('pinnacle' is the term used for the biological example identified) and abstraction phase and lightest light grey is the determination phase.

The building challenge(s) (e.g., heat gain, water gain and air exchange) is/are identified by the designer based on the expectations of the occupants and the environmental context.

The sub-phases in Figure 4-5 follow the flow of work as set out below:

1) Relevant biological challenges similar to those identified by the designers are investigated; in Figure 4-6 these are termed 'factors', 'processes' and 'pinnacles'.

2) A number of organisms (pinnacles) are chosen for further elaboration.

3) These pinnacles are analysed by the designer to distinguish their main features, strategies, mechanisms, and principles.

4) This is achieved by categorising the data produced by the analysis into specific groups.

5) The extracted data is abstracted and "imaginary pinnacles" are identified. For each function, the imaginary pinnacle holds the dominant or common feature of every individual category.

6) A design matrix is generated to produce design concepts.

Badarnah (2012) also developed a series of design tools to assist designers in the preliminary design phase. These are the exploration model, pinnacle analysis model, pinnacle analysis matrix and design path matrix. A brief explanation of each is provided below.

The exploration model is a hierarchical representation of biological knowledge and was developed as an example of a data mapping structure. This would be used by designers to narrow down their search from a general challenge to more specific aspects such as function, process and factors, so as to end with the relevant organism(s). The final biological examples are termed pinnacles.

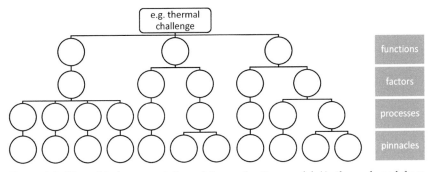

Figure 4-6. Hierarchical representation of the exploration model (Author, adapted from Badarnah (2012)).

What this process inherently means is that generation of the final design concept is directly affected by the exploration phase. This was noted by Badarnah (2012):

> "The current work provides a selection of representative processes and factors based on the analysis of a rather modest number of pinnacles, negligible compared to the sample size nature provides. In order to create a reliable generalized database, one needs to carry out extensive research on organisms and natural systems, which requires various resources and collaboration of professionals from numerous disciplines. Consideration of a wider number of pinnacles should result in a refined selection of optimized processes and factors. However, the sample size to be considered and the number of features/processes and their categorization remain a great challenge at this stage.
>
> Even when an extendable database becomes available, with the refined processes and factors, there would still be a need of a continuous investigation and update for such a database, since nature is continuously developing and updating." (Badarnah 2012)

The pinnacle analysis model is based on a series of steps by following which the main feature(s) of the performance of the pinnacle can be understood. This model analyses the pinnacle by investigating the strategy, the mechanism, the main principle and the main features of the thermoregulation carried out by the pinnacle organism. For example, Figure 4-7 shows the analysis of termite mounds.

The pinnacle analysis matrix is built on the pinnacle analysis model and is aimed at representing pinnacles by classifying the relevant data from the analysis in the previous step in a manner such that the proposed categories contain a potential analogy for the building envelope. These categories present the pinnacles by describing a set of thermal adaptation

Strategy	Mechanism	Main principles	Main features
The inhabitants of the mound modify it in accordance with environmental changes for homeostasis	Structural features to retain or dissipate heat: e.g. variations in wall thicknesses, surface pattern, projecting structures, orientation, chimneys, air passages, porosity	Natural convection	Chimneys and air passages

Figure 4-7. An example of the pinnacle analysis model (Author, adapted from Badarnah (2012)).

	Processes			Flow		Adaptation			Scale				Environmental context						Morphological features							Material features			Other features				
	a	b	c	a	b	a	b	c	a	b	c	d	a	b	c	d	e	f	a	b	c	d	e	f	g	a	b	c	a	b	c	d	e
Pinnacles	●				●		●			●			●	●	●	●					●			●		●				●			
	●				●		●			●				●							●						●		●		●		
		●		●			●			●			●	●	●	●						●					●				●		
Function — Imaginary pinnacle	●				●		●			●				●							●						●				●		

Figure 4-8. Pinnacle analysis matrix (Author, adapted from Badarnah (2012)).

characteristics. These are: the processes, flows (passive or active), different types of adaptation (physiological, morphological and behavioural), scales (nano, micro and macro), climatic classifications (e.g., arid, polar), morphological features, and material features (Figure 4-8).

The pinnacle analysis matrix was developed to reduce the complexity of the solutions a designer might face in the case where multiple relevant pinnacles are found and accordingly, multiple strategies, mechanisms, principles and features. Hence, for each function, an imaginary pinnacle is introduced that holds the dominant or common feature of every individual category. Once the imaginary pinnacles are established, the design path matrix is created to highlight the dominant features and thus facilitate the generation of design concepts (Figure 4-9).

Even though Badarnah (2012) referred to BioGen as a systematic design methodology by which relevant biological principles can be found, abstracted and then applied in the generation of a concept, the first step to finding the relevant thermoregulatory strategies in nature is not fully explored. This is because only a small number of organisms were investigated for creating the exploration model, which was then developed as a design tool for use in the preliminary biomimetic design phase. For this reason, the proposed biomimetic methodology cannot be said to have been generalised based on exploring all thermal adaptation

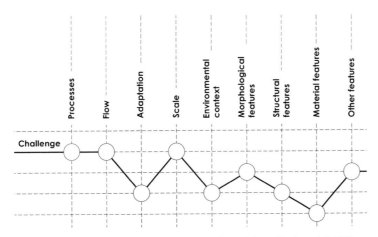

Figure 4-9. Design path matrix (Author, adapted from (Badarnah 2012)).

strategies in nature. Badarnah's research remains mainly focused on the transformation of strategies available in nature into technical solutions for building envelopes.

The design path matrix is the end stage of the preliminary design phase and the point at which the emulation phase starts. For each environmental challenge, and based on the dominant features, a design concept is developed and its energy performance then evaluated to ensure the bio-inspired building envelope is energy efficient. Later, Badarnah and Kadri (2015) refer to this method as a biophysical framework that can be used for generating building envelope design concepts. They state that the database of natural organisms provided for the proposed framework is not comprehensive in terms of introducing a generalised list of processes, factors and hence, pinnacles, which in turn implies the need for future exploration and elaboration. This book is part of that future exploration.

4.4 Assessment

This section looks at what can be learned from the examples discussed in this chapter and what next steps are needed to give designers access to all the thermoregulatory strategies used in the natural world.

4.4.1 The examples of biomimetic design

The examples discussed in this chapter confirm that biomimetic design is the focus of increasing attention. It is also clear that biomimicry is an applied science that derives inspiration from the natural world and that can open avenues for technological and sustainable design. As researchers delve more into nature, more sustainable solutions in terms of efficient processes, functions, systems and materials are being explored. Although much research has been conducted with the aim of systematising a process for bio-inspired design, as yet there is no database of biological thermoregulatory mechanisms. Although there are databases for assisting architects and engineers with biomimetic design, examples being AskNature (AskNature 2021), BioTRIZ (Gamage and Hyde 2011), SAPPhire (Chakrabarti et al. 2005) and IDEA-INSPIRE (Chakrabarti et al. 2005), none contribute directly to thermal issues by using thermal adaptation strategies found in nature. For example, even though AskNature to some degree supports a systematic search, it is a framework with a general purpose. This means the functional categories from which exploration of the biological database begin are not directly associated with thermal adaptation principles. BioTRIZ connects design problems to previous biomimetic solutions. Likewise, SAPPhIRE and IDEA-INSPIRE are data banks of previous biomimetic research and are mostly used by mechanical engineers.

Like other fields of knowledge, architectural design has benefited from biomimicry and the literature shows that the performance of some buildings has been enhanced by applying bio-inspired thermal adaptation strategies to building design. However, looking at the examples it seems most suggestions have imitated the morphological aspects of natural organisms. Most studies in this area have been metaphorical and thermoregulation was achieved by imitating the patterns, structures or functions of natural organisms, leading to the geometrical manipulation of building facades (Wang and Li 2010, Wang 2011, Park and Dave 2014). However, there has been an evolution in design approaches, as the design of a building skin based on bio-inspired principles has evolved from a simple imitation of the patterns and geometries found in nature and applying these to a façade, to designing specific types of wall systems or creating advanced materials (Alston 2015, Reichert et al. 2015). Additionally, of the 29 references given in Table 4-2, 65% were focused on just one organism and the associated thermoregulation strategies. This demonstrates no comprehensive research has yet been conducted in terms of generalizing biological principles.

Current knowledge of natural principles is also scattered. Biological thermal adaptation principles are also yet to be documented as a generalised dataset that would enable designers to connect existing thermal challenges to relevant biological solutions. Despite this there is some evidence that the energy performance of buildings has been enhanced by applying bio-inspired thermal adaptation strategies to building design (Tachouali and Taleb 2014, Al Amin and Taleb 2016, Zuazua-Ros et al. 2016). However, the focus has largely been on the building façade or envelope, and as yet there has been no consideration of the whole building as a living organism.

Most research in this area appears to be focussed on hot climates. The fact that more research has been conducted in hot climates suggests there could be insufficient information about thermal issues in cold climates, adaptation strategies for organisms in cold climates have not yet been investigated or even that thermal issues in cold climates are of less concern for researchers in this field. It might also be that hot climates are an easier starting point because the one-dimensional need to be cool is a simpler problem. However, in a colder climate, summer cooling and winter heating may be required, with gradations of environmental control in between.

Almost all research based on pre-existing knowledge about biological examples fails to refer to any specific process by which relevant information on thermal regulation strategies in nature can be found. Using natural systems to overcome thermal issues in buildings necessitates a systematic biomimetic design approach, and only the study by Badarnah (2012) has made a notable effort to develop an approach along these lines.

4.4.2 Badarnah's approach to biomimetic design

Badarnah claims the design tools she developed are flexible in their input and firm in their output. This implies that the more varied the challenges, and subsequently the more varied the functions, processes, factors and pinnacles, the more dominant features are found and hence, more imaginary pinnacles are extracted. Badarnah also acknowledges that despite the availability of myriad biological sources, finding pinnacles and abstracting their principles remains a big challenge for bio-inspired design and designers. She also points out the difficulty of managing a huge exploration model with a large quantity of pinnacles, as she believes the newly generated dominant features of one solution path might affect the solutions to other thermal challenges in the same building.

4.4.3 The next step

The development of the ThBA described in this book began at this point in the studies described above, and particularly in the work of Badarnah. The aim with the ThBA was to develop a comprehensive framework that would be generalisable and include all thermoregulatory mechanisms that could be identified in the natural world. How this was done and tested is explained in the remaining chapters of this book.

References

Ahmar, S. E. and Fioravanti, A. (2014a). *Botanics and Parametric Design Fusions for Performative Building Skins—An Application in Hot Climates.* Paper presented at the Fusion, Proceedings of the 32nd International Conference on Education and Research in Computer Aided Architectural Design in Europe, Newcastle upon Tyne, UK.

Ahmar, S. E. and Fioravanti, A. (2014b). *How Plants Regulate Heat, Biomimetic Inspirations for Building Skins.* Paper presented at the 48th International Conference of the Architectural Science Association 2014, Genoa, Italy.

Ahmar, S. E. and Fioravanti, A. (2015). *Biomimetic-Computational Design for Double Facades in Hot Climates—A Porous Folded Façade for Office Buildings.* Paper presented at the Proceedings of the 33rd eCAADe Conference, Vienna, Austria.

Al Amin, F. and Taleb, H. (2016). *Biomimicry Approach to Achieving Thermal Comfort in a Hot Climate.* Paper presented at the SBE16 Dubai, Dubai, UAE.

Alkhateeb, E. and Taleb, H. (2015). Redesigning of building envelope: tree bark as a biomimicry concept. *WIT Transactions on Ecology and the Environment, 193*(7), 1019–1029.

Alston, M. (2015). Natures buildings as trees: biologically inspired glass as an energy system. *Optics and Photonics Journal, 5*(136), 136–150.

AskNature. (2021). It's time to ask nature. Retrieved 12 October 2020, from The Biomimicry Institute—Nature-Inspired Innovation https://asknature.org/.

Badarnah, L. and Knaack, U. (2007). *Bio-Inspired Ventilating System for Building Envelopes.* Paper presented at the International Conference of 21st Century on Building Stock Activation.

Badarnah, L., Farchi, Y. N. and Knaack, U. (2010). Solutions from nature for building envelope thermoregulation. pp. 251–262. *In:* A. Carpi and C. Brebbia (eds.). *Design & Nature V: Comparing Design in Nature with Science and Engineering.* Southampton, UK: WIT Press.

Badarnah, L. (2012). *Towards the Living Envelope: Biomimetics for Building Envelope Adaptation.* (Doctoral dissertation), TU Delft, Zutphen, Netherlands.

Badarnah, L. (2015). A biophysical framework of heat regulation strategies for the design of biomimetic building envelopes. *Procedia Engineering, 118*, 1225–1235. doi:10.1016/j.proeng.2015.08.474.

Badarnah, L. and Kadri, U. (2015). A methodology for the generation of biomimetic design concepts. *Architectural Science Review, 58*(2), 120–133. doi:10.1080/00038628.2014.922458.

Bermejo-Busto, J., Martin Gomez, C., Zuazua Ros, A., María Ibáñez-Puy, M., Miranda Ferreiro, R. and Baquero Martin, E. (2016). Improvement of an integrated Peltier HVAC system integrated using beehive and stigmergy strategies. *DYNA Ingenieria E Industria, 91*(1), 507–511. doi:10.6036/7865.

Chakrabarti, A., Sarkar, P., Leelavathamma, B. and Nataraju, B. S. (2005). A functional representation for aiding biomimetic and artificial inspiration of new ideas. *AI EDAM-Artificial Intelligence for Engineering Design Analysis and Manufacturing, 19*(02), 113–132. doi:10.1017/s0890060405050109.

Elghawaby, M. (2010). *Biomimicry: A New Approach to Enhance the Efficiency of Natural Ventilation Systems in Hot Climate.* Paper presented at the Architecture and Research: The Third International Seminar-Arquitectonics Network, Barcelona, Spain.

Freepic. (n.d.). Honey bee with its silhouette on transparent background Free Vector. Retrieved from https://www.freepik.com/free-vector/honey-bee-with-its-silhouette-transparent-background_11207007.htm, Freepic License: 2010–2021 Freepik Company S.L.

Gamage, A. and Hyde, R. (2011). *Can Biomimicry, as an Approach, Enhance Ecologically Sustainable Design (ESD)?* Paper presented at the 45th Annual Conference of the Architectural Science Association: From Principles to Practice in Architectural Science, Sydney, Australia.

Gruber, P. and Gosztonyi, S. (2010). Skin in architecture: towards bioinspired facades. *Design and Nature V: Comparing Design in Nature with Science and Engineering, 138*, 503–513. doi:10.2495/Dn100451.

Han, Y., Taylor, J. E. and Pisello, A. L. (2015). Toward mitigating urban heat island effects: Investigating the thermal-energy impact of bio-inspired retro-reflective building envelopes in dense urban settings. *Energy and Buildings, 102*, 380–389. doi:10.1016/j.enbuild.2015.05.040.

Imhoff, M. L., Zhang, P., Wolfe, R. E. and Bounoua, L. (2010). Remote sensing of the urban heat island effect across biomes in the continental USA. *Remote Sensing of Environment, 114*(3), 504–513.

Kim, K. H. and Torres, A. (2015). *Integrated Facades for Building Energy Conservation*. Paper presented at the International Conference on Circuits and Systems (CAS 2015), Paris, France.

Lee, C. (2008). *The Thermal Organism and Architecture*. Paper presented at the Silicon + Skin: Biological Processes and Computation, Proceedings of the 28th Annual Conference of the Association for Computer Aided Design in Architecture (ACADIA), Minneapolis, MN.

Lopez, M., Croxford, B., Rubio, R., Martín, S. and Jackson, R. (2015). Active materials for adaptive architectural envelopes based on plant adaptation principles. *Journal of Facade Design and Engineering, 3*(1), 27–38.

Lopez, M., Rubio, R., Martín, S., Croxford, B. and Jackson, R. (2015). *Adaptive Architectural Envelopes for Temperature, Humidity, Carbon Dioxide and Light Control*. Paper presented at the 10th Conference on Advanced Building Skins, Bern, Switzerland.

López, M., Rubio, R., Martin, S., Croxford, B. and Jackson, R. (2015). Active materials for adaptive architectural envelopes based on plant adaptation principles. *Journal of Facade Design and Engineering, 3*(1).

Loughner, C. P., Allen, D. J., Zhang, D.-L., Pickering, K. E., Dickerson, R. R. and Landry, L. (2012). Roles of urban tree canopy and buildings in urban heat island effects: Parameterization and preliminary results. *Journal of Applied Meteorology and Climatology, 51*(10), 1775–1793.

Nanaa, Y. and Taleb, H. (2015). The lotus flower: biomimicry solutions in the built environment. *WIT Transactions on Ecology and the Environment, 193*, 1085–1093.

Nessim, M. A. (2015). Biomimetic architecture as a new approach for energy efficient buildings through smart building materials. *Journal of Green Building, 10*(4), 73–86. doi:10.3992/jgb.10.4.73.

Oke, T. R. (1973). City size and the urban heat island. *Atmospheric Environment, 7*(8), 769–779.

Park, J. J. and Dave, B. (2014). Bio-inspired parametric design of adaptive stadium facades. *Australasian Journal of Construction Economics and Building Conference Series, 2*, 27–35.

Pixabay. (2017). Beehive. Retrieved from https://pixabay.com/vectors/bee-beehive-hive-honey-honeycomb-2022492/, Pixabay License: https://pixabay.com/service/license/.

Reddi, S., Jain, A. K., Yun, H.-B. and Reddi, L. N. (2012). Biomimetics of stabilized earth construction: Challenges and opportunities. *Energy and Buildings, 55*, 452–458. doi:10.1016/j.enbuild.2012.09.024.

Reddy, T. A., Maor, I. and Panjapornpon, C. (2007). Calibrating detailed building energy simulation programs with measured data. *HVAC&R Research, 13*(2), 221–241. doi:10.10 80/10789669.2007.10390952.

Reichert, S., Menges, A. and Correa, D. (2015). Meteorosensitive architecture: Biomimetic building skins based on materially embedded and hygroscopically enabled responsiveness. *Computer-Aided Design, 60*, 50–69. doi:10.1016/j.cad.2014.02.010.

Sara, K. and Noureddine, Z. (2015). *A Bio Problem-Solver for Supporting the Design, Towards the Optimization of the Energy Efficiency*. Paper presented at the 6th International Conference on Modeling, Simulation, and Applied Optimization (ICMSAO), Istanbul, Turkey.

Scartezzini, J. L., Martín-Gómez, C., Bermejo-Busto, J., Zuazua-Ros, A., Miranda, R. and Baquero, E. (2015). *Redesign of the Integration of Building Energy from Metabolisms of Animal: The RIMA Project.* Paper presented at the International Conference CISBAT - Future Buildings and Districts, Sustainability from Nano to Urban Scale, Lausanne, Switzerland.

Siegel, A., Hui, J., Johnson, R. and Starks, T. P. (2005). Honey bee workers as mobile insulating units. *Insectes Sociaux, 52*(3), 242–246.

Tachouali, D. M. and Taleb, H. M. (2014). *Flamingo Strategies in Approaching Sustainable Design in Built Environment: Case Study in Dubai.* Paper presented at the International Conference on Architecture And Civil Engineering (ICAACE'14) Dubai, UAE.

Turner, J. S. and Soar, R. C. (2008). *Beyond Biomimicry: What Termites can tell us About Realizing the Living Building.* Paper presented at the First International Conference on Industrialized, Intelligent Construction, Loughborough, UK.

Wang, J. and Li, J. (2010). *Bio-inspired Kinetic Envelopes for Building Energy Efficiency based on Parametric Design of Building Information Modeling.* Paper presented at the Asia-Pacific Power and Energy Engineering Conference, Chengdu, China.

Wang, J. (2011). Bio-inspired Kinetic Envelopes: Integrating BIM into Biomimicry for Sustainable Design. Retrieved 19 June 2018, from Boston Society of Architects (BSA) https://www.architects.org/sites/default/files/NEW_Report-Julian%20Wang.pdf.

Worall, M. (2011). Homeostasis in nature: Nest building termites and intelligent buildings. *Intelligent Buildings International, 3*(2), 87–95. doi:10.1080/17508975.2011.582316.

Zare, M. and Falahat, M. (2013). Characteristics of reptiles as a model for bionic architecture. *Advances in Civil and Environmental Engineering, 1*(3), 124–135.

Zuazua-Ros, A., Martín-Gómez, C., Bermejo-Busto, J., Vidaurre-Arbizu, M., Baquero, E. and Miranda, R. (2016). Thermal energy performance in working-spaces from biomorphic models: The tuna case in an office building. *Building Simulation, 9*(3), 347–357. doi:10.1007/s12273-016-0273-8.

Chapter 5
Developing a Structure for the ThBA

5.1 Environmental adaptation: a leap forward for energy efficiency

In his discussion of the concepts underlying sustainable development, Jabareen (2008) details the concept of 'eco-form' that "...represents the ecologically desired form and design of the human habitat such as urban spaces, buildings, and houses." Waheed et al. (2009) take this further, seeing eco-form as the "...ecologically desired spatial form of cities, villages, and neighbourhoods." Quite what such a spatial form might look like is not described, although the eco-form approach is believed to play an important role in climate change mitigation through the optimisation of the spatial properties of human settlements (Jabareen 2013). In this context, understanding how natural organisms adapt to their environment could represent a practical approach to ecological design (Van der Ryn and Cowan 2013). The habitat modifications that animals and birds create for themselves, such as nests and burrows are related to climate and the natural materials at hand, as well as being bio-degradable; all properties that an eco-form might need to imitate. It therefore seems important to study living things as "the world is a vast repository of unknown biological strategies that could have immense relevance should we develop a science of integrating the stories embedded in nature into the system we design to sustain us" (Todd and Todd 1994).

In a similar but functional-oriented approach, an understanding of the functions and processes found in ecosystems has led to the development of environmental theories some of which would seem to have had created the foundations for biomimetic energy-efficient building design (Gamage and Hyde 2011). As an example, the Self-Organising Holarchical Open System (SOHO) was formed to describe the relationship between natural and man-made structures (Waltner-Toews et al. 2008). SOHO was thus a conceptual model that sought to bring together ecological integrity and human sustainability. In this model, both human and natural ecological systems are perceived as entities within a larger environmental context. Of interest for this book is the fact that these systems can be seen as buildings (human systems) and biological organisms (natural systems) that are both affected by their surrounding environments. Consequently, the SOHO model was developed to help designers understand "ecosystems as complex self-organizing hierarchical open systems that adapt to their environment" (Kay 2003).

A living organism is part of a larger encompassing system—known as an ecosystem—in which everything uses a recursive flow of materials and energy with the one additional input of solar energy. Organisms receive energy from the solar charged ecosystem, process and use it for their sustenance and finally, return some of it back to the environment. This cycle can be understood through non-equilibrium thermodynamics in which the properties of a system change as energy dissipates during a transformation from one state to another (Muller 2000). This is similar to buildings as these consume primary energy sources both for their construction, maintenance and operation and could return part of the input energy to the environment at the demolition stage. It appears that, in both biological and man-made systems, the energy needed for the system to function could be reduced when the system is fully integrated into its larger context. If more use could be made of demolition materials and the energy contained in these in place of using virgin materials for new buildings, energy would be saved.

As discussed in Chapter 1, the overarching aim of Ecologically Sustainable Development (ESD) is to develop strategies for climate adaptation and ultimately to reduce the climate change impacts of a building on the environment (GhaffarianHoseini 2012). As shown in Chapter 2, vernacular building traditions were highly influenced by climate, and considering climatic conditions in modern building design can make a significant contribution in reducing operational energy (Cheung et al. 2005, Holmes and Hacker 2007). Buildings are expected to maintain a thermally comfortable indoor environment irrespective of the external climatic conditions, and currently the energy needed for this is mainly gained from burning fossil fuels. Given these points, in the context of ESD, energy-efficient buildings can be defined as those with enhanced thermal performance, achieved through the adaptation of the building to its environment in order to create climatically appropriate thermally comfortable conditions within. As a result, understanding the principles of the adaptations used by organisms might suggest ideas for innovative energy-efficient building design. Similar to a living organism which uses different strategies to regulate its body temperature to survive in a harsh climate, more than one strategy might be needed in order to maintain the internal temperature of a building within an appropriate comfort range in the face of fluctuating external temperatures. Currently, most buildings show static behaviour with regard to interacting with their environments, which does not lead to energy efficiency (Fernández et al. 2013, Loonen et al. 2017).

As discussed in 3.1.3.2.2, the translation of biological thermoregulatory principles into architectural design has led to reduced energy use in buildings. Therefore, it seems essential to investigate how to find relevant biological thermoregulatory solutions either to solve a pre-identified thermal challenge or to act as a starting point for design.

5.2 Literature review

As discussed in Chapter 4, previous research showed the thermal adaptation solutions used by animals and plants have not yet been collected together in a systematic way. It also appears that the list of organisms and their associated biological thermoregulatory solutions is incomplete. Therefore, at the start of this research a comprehensive literature review was conducted to create an inclusive list of biological thermal adaptation strategies. The aim was to classify and generalise these as grouping the strategies could facilitate future use of them by architects.

The categorisation scheme used in this book is the result of the three steps of the literature review (Figure 5-1).

1) The first step explains the methods of biological heat transfer.

2) This is followed by introducing the existing non-thermoregulatory systems in biological science which have been developed in order to classify all types of living organisms. In biological science, the term 'thermal adaptation' is usually associated with the behaviour, morphology, or physiology of an organism. Therefore, exploring morphological, behavioural and physiological thermal adaptation strategies was also considered in the second step of the literature review. The hope was that thermal adaptation strategies would already be grouped under these categories.

3) Studying the thermal physiology of heat regulation was the third step. The purpose was to assist in creating a foundation for categorising the thermal adaptation strategies. In this context, the glossary of thermal physiology terms was used as a reference. This was produced by the International Union of Physiological Sciences (IUPS) and had undergone many reviews by experts during its consultative development, which subsequently resulted in several iterations.

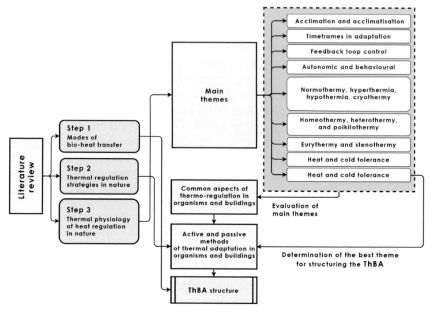

Figure 5-1. Steps in the literature review and their contribution to creating the final hierarchical classification of biological thermal adaptation strategies (Author).

This book used Version 3 which includes 497 terms describing different aspects of thermal adaptation mechanisms. It is essential to use the terms on which the experts have agreed and therefore, where they appear in this book, they are quoted in full to minimise misinterpretation.

5.2.1 Step 1: basics of bio-heat transfer

Thermoregulation in nature happens through dry and wet heat transfer. The former is also called dry heat exchange or sensible heat transfer (Bakken 1976) and occurs through conduction, convection and radiation, while in the latter, heat loss occurs through the evaporation of water. In addition to wet and dry heat transfer, to analyse the uptake and use of energy by living organisms, the third parameter of heat generation in the body of the organism needs to be considered.

The rate and direction of conductive heat transfer depends on the temperature difference between the body of an animal and the surface with which the body is in contact. For some species of large mammals, the rate and direction of conductive heat transfer can be adjusted through the use of a specific behavioural thermoregulation strategy. For example, Chacma baboons (*Papio Hamadryas Ursinus*) who live in the Namib Desert use sand bathing (Brain and Mitchell 1999) to increase the conductive heat transfer between their body and the cooler sand beneath them, which leads to a reduction in their body temperature.

Convection is a distinct mode of heat transfer that occurs through a current of moving gas or fluid surrounding the body of the animal. It happens due to the temperature difference between the body of an animal and its immediate microclimate, which is itself influenced by the surrounding air in the location where the animals live (Huey et al. 2012, Varner and Dearing 2014, Pincebourde et al. 2016). In marine animals, water is the medium of heat transfer while for land-based species, convective heat transfer happens through the air.

For air, in 'forced' convective heat exchange the wind blowing will carry the heat away, while for free convective heat exchange there would be no wind to enhance heat transfer. The high-speed movement of an animal will force convective heat transfer but at the same time, muscular activity generates heat so a thermal balance needs to be achieved.

Heat can also be transferred by radiation, an example being the heat absorbed by the body of an animal lying in the sun. Even though conductive and convective heat transfer play important roles in thermoregulation, radiative heat transfer is the most significant mode of heat exchange in the bodies of large mammals (Mitchell et al. 2018). Thermo-regulation occurs through animals seeking shade (Mole et al. 2016), or changing posture or orientation (Maloney et al. 2005) as means of adjusting the heat the animal body parts receive through radiation.

5.2.2 Step 2: classification measures of biological thermal regulation strategies

The literature suggests several non-thermoregulatory classification systems for categorising living organisms. However, looking at the species grouped under these classification systems, no similarities can be found in their thermal adaption strategies. While, the existing systems are not effective enough to inform the classification of thermal adaptation strategies, some of the measures that have been used to group animals and plants are explained here as these affect the ways in which animals and plants thermoregulate. These are: (1) taxonomic categories, (2) the climatic conditions of organism habitats and (3) the different scales at which thermal adaptation takes place (Figure 5-2).

The taxonomic classification system of living organisms has eight main levels from the most inclusive category at the top, to the most exclusive at the bottom. These are domain, kingdom, phylum, class, order, family, genus and species. Each level splits into more specific groups. Kingdom is the highest level under which all organisms are classed. For example, the six recognised main kingdoms in nature are plants, animals,

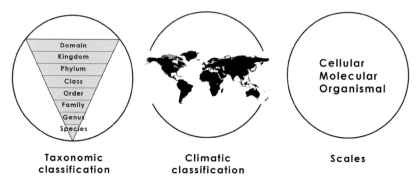

Figure 5-2. Existing non-thermoregulatory classification systems of living organisms (Author, adapted from Pixabay (n.d-b)).

bacteria, archaebacterial, fungi and protozoa. The categories presented by the phylum level are created to show major differences in the physical characteristics of organisms as each contain closely related animals and plants. From there to the species level, the same pattern is used to split organisms in each individual level. Species is the most specific taxonomic rank where organisms are given a specific name that is unique only to them.

Looking at the hierarchical classification of animals and plants, each level contains categories of organisms that share similarities in physical characteristics but none use similar thermoregulatory solutions.

In conducting the literature review for this research it seemed it would be possible to group plants into distinctive categories depending on how they respond to water level and light intensity, both factors influenced by the location and hence climate in which a plant grows (Figure 5-3). The physiological and morphological characteristics that plants have evolved are influenced by the prevailing environmental stimuli in their habitants.

'Spiral phyllotaxis' and 'distichous phyllotaxis' are the two types of evolutionary form plants have evolved in response to light intensity.

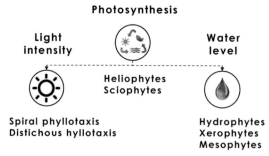

Figure 5-3. Types of plants based on their response to environmental stimuli (Author).

Phyllotaxis is defined as "the regular arrangement of lateral organs around a central axis" (Smith et al. 2006). The presence of the Fibonacci sequence in the arrangements of the leaves is called 'Spiral phyllotaxis'. It is the most prevailing pattern found in plant leaves and is assumed to maximise the light absorption while also improving diffusion (Valladares and Brites 2004) (Figure 5-4).

In another approach, plants are divided into three categories relative to their access to water. Plants surrounded by water are called hydrophytes. Mesophytes do not grow near water but have access to it so they can absorb enough when it is required. Compared to hydrophytes and mesophytes, xerophytes experience water deficit and have thus evolved to store water to survive.

Even though the classification systems mentioned above present a clear structure for explaining how plants respond to water and light levels, none directly explain how they regulate temperature. For example, it is not clear whether the thermal adaptation strategies of 'distichous phyllotaxis' plants are the same, or whether adaptation to light by mesophytes that happens through changes in their stomata size is similar to ways in which they adapt to temperature. However, it seems plausible to link light and temperature as there is a direct correlation between them, since a higher light intensity leads to an increase in temperature. This could possibly help in classifying thermal adaptation solutions.

Similarly, when it comes to the animal kingdom, the literature does not confirm whether all animals in cold climates use similar thermal adaptation strategies; for example, some form of counterflow heat exchange to avoid their extremities freezing.

To perform photosynthesis, plants employ different strategies to adjust the uptake of light and carbon dioxide while keeping water transpiration from the leaves at an optimised level. Subsequently, the irradiance and hydrothermal conditions of the climates in which plants grow have led to a third categorisation scheme that groups plants into heliophytes and sciophytes. Similar to the previous classification schemes, the categorisation of plants based on the photosynthesis method they use was unhelpful for classifying thermoregulatory strategies.

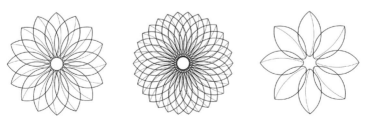

Figure 5-4. The three different divergence angles between leaves with spiral phyllotaxis patterns (Author, adapted from (Valladares and Brites 2004)).

Regarding categorisation based on scales, the existing research does not organise thermoregulatory solutions based on the levels where they take place in the bodies of animal or plants. Thermoregulatory mechanisms can be seen either at the sub-cellular, cellular or organ level, but no mention is made of a relationship between these.

Overall, step two produced examples but it was also obvious that thermoregulation strategies differed in the groups formed by the three different types of classification. For example, mammals living in cold climates use different thermoregulatory solutions even if they belong to the same family (Figure 5-5). The arctic fox uses countercurrent heat exchange in its legs, polar bears have thick pads and fur on the feet to keep these warm, while other bears hibernate to survive the cold winter. As step two did not seem to help in organising the available strategies, it was abandoned and step three started.

Figure 5-5. The taxonomy classification of grizzly bear (Author, adapted from (Dupont 2009, Davidvraju 2013, Airwolfhound 2015, Gzen92 2019, Anderson 2019, Beaufort n.d., Pixabay n.d-a)).

5.2.3 Step 3: thermal physiology of heat regulation in nature

Understanding the themes discussed in the glossary of thermal physiology was necessary for providing a comprehensive perspective on the main concepts of thermal adaptation. The results of step three of the literature review are explained below as a connection emerged between the findings that enabled the biological thermoregulatory solutions to be structured.

The themes discussed below are different from the thermal adaptation strategies which will be discussed in Chapter 6. The concepts reviewed here are done so in the hope of suggesting a structure for classifying thermal adaptation strategies. This has been referred to as a 'classification scheme' throughout the text.

5.2.3.1 Endothermy and ectothermy

Everyone is familiar with the idea of warm- and cold-blooded animals. The former, including people, keep their bodies at a more or less constant temperature, whereas the latter, including lizards and snakes, need to warm themselves in the sun to increase their body temperatures, or seek shade to cool the same. Warm-blooded animals are officially known as endotherms and they can generate heat within their bodies, whereas the body temperature of ectotherms, the official term for cold-blooded animals, changes with the temperature of the environment. The process by which heat is generated within the body is known as thermogenesis. All mammals and birds are endotherms and generate heat through their normal metabolic activity. Although insects are generally considered to be ectotherms, some benefit from muscular thermogenesis to generate internal heat (Davenport 2012). The spruce budworm is an insect that benefits from diet-induced thermogenesis (Trier and Mattson 2003). However, heat generation is not found in most ectotherms and thus they face thermal challenges in the event of changes in the ambient temperature. The metabolic rate in their bodies is such that they usually use behavioural thermoregulation strategies to adapt to their environment, such as basking in the sun. In contrast, the high rate of metabolic heat generation enables endotherms to keep a stable internal body temperature. However, endotherms, like ectotherms, still exchange heat with the surrounding environment and this leads to adaptation so the body becomes used to maintaining a constant body temperature whether in a hot or a cold climate. The terms used to describe this adaptation are called acclimation and acclimatisation.

5.2.3.2 Acclimation and acclimatisation

The International Union of Physiological Science (IUPS) state acclimation is the "Physiological or behavioural changes occurring within an organism, which reduce the strain or enhance endurance of strain caused by experimentally induced stressful changes in particular climatic factors". Thermal adaptation in this term occurs in response to induced changes in a specific climatic factor such as ambient temperature (Eagan 1963). Acclimatisation is defined as the "Physiological or behavioural changes occurring within the lifetime of an organism that reduce the strain caused by stressful changes in the natural climate (e.g., seasonal or geographical)" (IUPS Thermal Commission 1987). Many dictionaries give the same definition for acclimation and acclimatisation, apart from noting that the latter is the verb form. People acclimatise when they move from one climate to another and this is a form of adaptation, known as phenotypic adaptation, because it happens within the lifetime of an organism (Taylor 2006). As an example, a person moving to a high latitude where the level of oxygen in the environment is much lower will adapt in the short term "...because their bodies raise their levels of haemoglobin, a protein that transports oxygen in the blood. However, continuously high levels of haemoglobin are dangerous, so increased haemoglobin levels are not a good solution to high-altitude survival in the long term" (National Geographic 2020). Despite this, Tibetans live at high altitudes without this problem because of a genetic mutation, which allows them to use the available oxygen more efficiently. This type of adaptation is called genotypic adaptation and can happen over centuries.

Another example of genotypic adaptation is the peppered moth *Biston betularia*. The population of these moths had increased in Britain by the end of Industrial Revolution as they evolved dark-coloured wings, which enabled them to be camouflaged when settling on tree bark that had been darkened by the sooty atmosphere and thus escape predation by birds. The process of such an adaptation occurs through an eventual change in the gene frequency relative to the advantages conferred by a particular characteristic, in this case the wing coloration of the moths (Brakefield 1990, Majerus 2009). In some animals, the form of a feature has evolved by natural selection to enable a specific function, for example, the light bones of flying birds.

5.2.3.3 *Timeframes in adaptation: biorhythms*

Short time frames form another important aspect of adaptation generally associated with physiological and behavioural traits. From this perspective, the two types of adaptive activity known as crepuscular and nycthemeral describe the times during which animals use an adaptation strategy. Those which take place at dusk or dawn are called crepuscular while nycthemeral adaptation activities occur on a 24-hour basis. The latter is also referred to as the light/dark cycle (LD cycle).

Additional terms have been introduced to define a range of activity patterns that occur in the LD cycle. If the majority of the adaption activity happens at night then the pattern is nocturnal, while if it mostly occurs during the day or the light portion of the LD cycle, it is diurnal. The activity rhythms are ultradian when their period length is significantly short. Acyclic is an arrhythmic adaptive pattern for which no obvious pattern is observed (Halle and Stenseth 2012). For example, it has been argued that salmon migrate during the night because of the presence of predators during the day. This suggests external factors can affect the biorhythmic behaviour of animals; in this case, the swimming activity of salmon is impacted by their predators (Martin et al. 2012).

5.2.3.4 Feedback loop control

Homeostasis is the term used for the capability of organisms to stabilise the internal state of their bodies through responding to environmental stressors that are induced by a changing variable. The glossary of thermal physiology identifies the negative-feedback loop as the underlying process of homeostasis within which an organism is capable of maintaining equilibrium.

In biological systems, feedback control loops are the products of molecular and physiological processes which can be either simple or complex (Figure 5-6) depending on how the organism responds to thermal stressors. The three main components of feedback systems are the stimulus, sensor and effector (Waterhouse 2013). The former disturbs an originally controlled variable which will then be detected by the receptor. The message is subsequently sent by the receptor to the controller. The control centre determines the appropriate response and sends it to the effector to perform the action. In a homeostasis thermal balance mechanism, a sensor is responsible for monitoring variables. Responses to thermal stressors can be either physiological or behavioural and have been viewed as reflexes (Woods and Ramsay 2007).

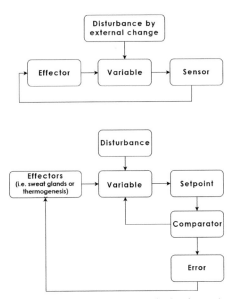

Figure 5-6. Principles of feedback control in a simple (at the top) and complex (on the bottom) feedback system adapted from Willmer et al. (2009).

In a negative feedback control system, the comparator analyses the signal received and compares it against a predetermined set point before the next step of thermal adjustment occurs. In this situation, the feedback loop is open as the adjustment process repeats until the received signal (such as the temperature) reaches the set point (Figure 5-6). For example, during sweating, the error sensors measuring hypothalamic temperature are in the bloodstream but the disturbance detectors are embedded in the skin. The latter are called 'cutaneous thermoreceptors' and these send data to the hypothalamus where the temperature is compared against a set point (Waterhouse 2013). The hypothalamus is a small organ at the base of the brain that regulates body temperature. In contrast, a positive feedback control system has a closed loop as the effector produces the same responses irrespective of being compared with a set point. An example of positive feedback in animals is the production of milk. As the calf takes the milk, a message is sent to the hypothalamus to release the prolactin hormone to increase milk production, so while the calf keeps sucking or the cow is milked by a machine, the cow will keep on producing milk.

Although issues like adaptation and feedback are part of the natural world, they have hitherto not provided ways of structuring the search for thermoregulatory systems that could provide inspiration for designers.

5.2.3.5 Autonomic and behavioural thermoregulation

The IUPS suggests different ways of classifying thermal regulation mechanisms, although some are obsolete and have been recently replaced by new classification terms. Potentially useful new terms are autonomic (physiological) and behavioural thermoregulation. In the former, the regulation of body temperature occurs through involuntary responses to thermal stressors with examples being physiological mechanisms such as sweating, shivering and non-shivering thermogenesis. The latter denotes an increase in heat production through metabolic activity rather than heat production through muscle activity (shivering). In contrast, behavioural thermoregulation takes place through voluntary movement of the organisms towards a thermally favourable environment where thermal conditions are appropriate for heat exchange.

Compared to animals, plants have less freedom in terms of moving in space, although plants can bend towards a heat source to maximise heat absorption, an action known as thermotropism. However, the IUPS does not recognise this as a thermoregulatory behaviour. Overall, it seems there is no well-defined distinction between thermotropism and thermoregulatory behaviour in plants. What is defined as behaviour in plants is limited to specific behavioural patterns such as "...foraging for light, nutrients, and water by placing organs where they can most efficiently harvest these resources" (Karban 2008). It has been argued that the response of plants to environmental stimuli is cognitive and intelligent enough to be counted as analogous to the behaviour of animals (Segundo-Ortin and Calvo 2019, Taiz et al. 2019).

To be able to structure biological thermal regulatory strategies it was necessary to reflect on the existing classification schemes as suggested by the IUPS. Reviewing the glossary of thermal physiology, the following grouping schemes were identified as fundamental in representing animals and plants based on the characteristics, limitations and scope of the responses they show to thermal stressors.

5.2.3.6 *Normothermy, hyperthermia, hypothermia, cryothermy*

The concept of normothermy, also known as cenothermy and euthermy, describe the body temperature being within its normal range (Clarke 2017). Hyperthermia points to a condition in which the core temperature rises to exceed the normal limit, while hypothermia happens when the core temperature falls below the normal range. The normal range of body temperature is, however, different for individual species, so that the minimum and maximum temperatures within which an organism can be in an active state vary.

The thermal status of an organism in a supercooling condition is known as cryothermy. In this condition, the body temperature drops below the freezing point of the body tissue.

5.2.3.7 *Homeothermy, heterothermy and poikilothermy*

The two concepts of homeothermy (homoiothermy) and poikilothermy describe the respective absence and presence of temperature variation in the body of an organism. The former is a thermoregulatory pattern in which the body temperature of a living organism remains constant regardless of changes in external temperature through use of metabolic energy. In more precise terms, the body temperature of homeotherms is comparatively constant as the changes occur only within a narrow range (Ivanov 2006). Homeothermy is associated with the high (tachymetabolic) and controlled rate of heat production in the body of an organism (Tzschentke and Rumpf 2011), in which the temperature variation usually takes place during nychthemeral or seasonal cycles. Homeotherms thus seem similar to endotherms but in fact, homeotherms are not necessarily endotherms. For example, some turtles keep their internal body temperature constant through behavioural strategies although they are not capable of heat generation in the manner of endotherms.

The thermoregulation pattern is known as heterothermy if the changes in the body temperature are greater than that of homeothermy. Regional or local heterothermy (Nord and Folkow 2018) is a particular type of heterothermy in which the temperature fluctuation takes place in specific regional tissues (Katz 2002). This means that in the body of such an organism, the temperatures of the thermal zones are different. This type of thermoregulation has been widely reported for endotherms living in cold climates (Brown and Baust 1980) and has been mostly observed in boreal, aquatic or subarctic mammals and birds (Johansen 1969). For example, the poor insulation in the wings of bats allows for more heat dissipation during flight when the ambient temperature is high while in cold weather, the wing and air temperatures remain relatively close (Rummel et al. 2019). Temporal heterothermy is another pattern in which the body temperature of an organism changes over time (Lovegrove 2000, Riek and Geiser 2014, Stawski et al. 2014); for example, animals that hibernate.

In contrast to homeothermy, the temperature variations in the body of poikilotherms occur over a broad range as they do not use metabolic energy for thermoregulation. These are the commonly called cold-blooded animals as the term poikilotherm is generally applied to vertebrates. Essentially the body temperature of a poikilotherm will vary, while an ectotherm living in a constant temperature environment will have a stable internal body temperature even though it has no internal heat generation mechanism, so it would be classed as a homeotherm. Conversely, while autonomic thermal regulation mechanisms are not generally used by poikilotherms, there are exceptions in some species. For example, large crocodilians and turtles use temporary muscular activity as a means of generating heat.

While the definition of homeothermy and poikilothermy is clear, it has been argued that heterothermy has no robust description as it includes all stages of thermoregulation between homeothermy and poikilothermy (Hensel et al. 1973).

5.2.3.8 Eurythermy and stenothermy

The IUPS glossary identifies the two distinctive groups of eurythermy and stenothermy to explain the extent to which an organism tolerates the temperature variations in its environment. The former shows that an organism can tolerate a wide range of environmental temperatures, something that happens with homeotherms (see 5.3.6). The latter describes the tolerance of organisms when the difference between maximum and minimum temperatures within which they can effectively operate is comparatively small (Moore 1940). Examples of the latter include heterotherms and poikilotherms, as the temperature variations in their bodies occur within a narrow range. Eurythermal and stenothermal strategies are also known as 'thermal specialist' and 'thermal generalist' (Logan and Buckley 2015). For example, most mammals and birds are thermal specialists as they are capable of thermal regulation through behavioural and physiological mechanisms (Angilletta Jr and Angilletta 2009).

Eurytherms can live in different types of habitats while stenotherms can only survive in a comparatively limited number of environments where the fluctuation of the environmental temperature is small. Ectotherms also tend to be localised to a particular climate and there is no single ectotherm capable of living in all climate types from the polar to the tropical (Pörtner and Gutt 2016).

Most organisms in the biosphere are poikilothermic or ectotherms (Adey and Loveland 2011). Given that, stenotherms usually live in cold environments while eurytherms are more abundant in tropical climates. In light of this, climate change effects can be destructive for the stenotherm species living in cold climates while eurytherms living in tropical regions can adapt to a wider range of temperatures.

5.2.3.9 Heat and cold tolerance

The IUPS describes cold tolerance, which is also known as cold endurance, as the ability of an organism to tolerate low ambient temperatures through a variety of physiological properties (IUPS Thermal Commission 1987). The IUPS also recognises certain types of homeotherms can endure cold environments either through insulation or efficient metabolic heat production. In glacial environments, thermal tolerance is achieved through local heterothermy and temporal heterothermy (5.2.3.7). In the former, organisms protect their appendages from freezing through vascular control, while an example of the latter is hibernation during which animals avoid the cold time by sleeping and temporarily stopping normal activities. Poikilotherms are also cold tolerant as some can endure freezing environmental temperatures using thermoregulatory biochemical reactions happening at molecular levels.

The IUPS defines heat tolerance (heat endurance) as the ability of an organism to tolerate high ambient temperatures, which like cold tolerance, involve a variety of physiological thermal adaptation strategies. For organisms to be called heat tolerant, they should be able to function normally when their body temperature exceeds the normal range. The IUPS identifies homeotherms as heat tolerant species. In contrast, poikilotherms are not counted as being heat tolerant as high temperatures in their bodies result in a type of death known as 'heat death'. However, the heat death points are not fixed and vary for different species (Bowler 2018), as these are significantly influenced by the environment to which an organism acclimates. There is evidence that warm acclimation increases and cold acclimation decreases the heat death points respectively (Cossins 2012).

Using another classification system developed for thermal adaptation strategies, organisms can respond to thermal stressors by avoidance, conformity or regulation (Willmer et al. 2009). To avoid thermal stresses, an organism gets away from the environment either by avoiding the space (an example being migration), or avoiding the time by stopping normal activities during a specific period, an example being hibernation (Willmer et al. 2009) (see 6.1.3, avoiding heat loss).

Thermal stresses mean the internal state of the body of an organism is affected by the external environment. For example, poikilotherms are temperature 'conformers' as their body temperature follows that of the environment. When responding to extreme thermal stressors, these species undergo a series of changes occurring at the biochemical level. Temperature conformity happens in the absence of effective physiological and behavioural thermal adaptation mechanisms. The biochemical changes enable organisms to stay functional with a minimal level of activity. An example is found in the potential damaging effects of freezing. To avoid these, an organism uses the thermotolerance or heat shock response (HSR). HSR is a short and rapid action taking place at a molecular level to protect cells from freezing and thus, to allow survival for a limited period within which an organism can retain normal activities.

Compared to thermal avoidance and the conformity mechanism, thermal regulation requires significant changes in an organism, which are usually a combination of behaviours and substantial physiological and chemical transformations.

5.2.4 Similar patterns of thermoregulation in organisms and buildings

Reviewing the themes discussed in the context of the thermal physiology of living organisms, it seems there are few similarities between the ways in which animals and organisms regulate temperature. The review was done in the hope it would reveal a way these themes could be classified. However, it might also be useful to discuss how they are different from what generally happens in buildings, with the aim that this might suggest a classification scheme.

As discussed earlier, there are two types of adaptive activities in living organisms—the crepuscular and nycthemeral (Grimpo et al. 2013), with the former happening only at dawn and dusk while the latter can occur anytime during the day. The nycthemeral patterns of behavioural and physiological functions in mammals are influenced by environmental stimuli, the timing system of the body of the organism and the ways in which this internal circadian system interacts with the environmental signals (Delmar Cerutti et al. 2019).

In a building, run-time patterns of HVAC systems are influenced by fluctuations in the temperature of the environment and the consistency of such patterns might lead to a fixed schedule for their automatic running. In addition, the running hours of some HVAC systems can be controlled manually by the occupants in immediate response to the local environmental conditions. HVAC running is also affected by the number of hours a building is occupied. In a building without a HVAC system, when the internal temperature is not within the comfort range, the direct interaction of the occupants with the space can lead to thermal regulation. For example, in such a building in a hot and humid climate, people might open the windows any time during the day to regulate the internal humidity and temperature.

It thus seems that automatic and user-activated running of the HVAC systems, and the direct interaction of users with internal space, occur around a range of temperature within which people feel thermally comfortable. This range is equivalent to the set-point in the body of animals. The running of the HVAC systems (both automatic and user-activated) can be seen as parallel to the physiological thermoregulatory strategies living organisms use to adapt to their environments while the direct interaction of users with internal space is more equivalent to a behavioural adaptive activity used by an organism. All these strategies have nycthemeral patterns as they can take place at any time during the day. However, in the automatic running of the HVAC system, the run-time schedule is influenced by the outside temperature for which maximum and minimum variations take place at certain times of the day. In this sense, cooling or heating systems turn on when there are higher temperatures midday and early afternoon and lower temperatures in the early morning and late evening. This is similar to the crepuscular pattern of biological thermoregulation happening at dusk and dawn. That being said, this analogy does not seem to be valid for offices or schools as the occupancy schedule of such spaces is less likely to happen at dawn or in the evening. This means the building type affects the rhythms of thermal regulation. The fact that the core temperature in humans changes during the day, due to the nycthemeral cycle of the physiological thermoregulation mechanisms happening in the body (Cabanac et al. 1976), does not seem to have a parallel in building thermoregulation.

The concept of the feedback loop in the body of organisms appears similar to the temperature control mechanisms and the thermoregulatory feedback found in HVAC systems but this link seems too general to form a classification scheme for organising the biological thermoregulatory strategies. The concept of feedback control appears to link to the concepts of normothermy, hyperthermia and hypothermia, which describe where the body temperature stands respectively regarding whether it is within, above or below the normal range in which an organism typically functions. These temperature states were seen as analogous to the fluctuating temperature in free running buildings. Also, the notion of a constant internal temperature in a building could be linked to the concept of homeothermy, while a fluctuating internal temperature could be linked to heterothermy, and poikilothermy. However, such a vague link does not suggest a classification scheme. These states either describe a range of temperature fluctuations or deviations from the set temperature and in all of them, the feedback control loop is the mechanism involved in regulating the temperature in the body of an organism or a building.

Similarly, the concepts of heat and cold tolerance at first had potential as a means of categorising thermal adaption strategies but then appeared inappropriate. Reviewing the three sub-branches of temperature tolerance—namely temperature avoidance, conformity, and regulation—biological thermal adaptation strategies could be grouped based on the cold and heat endurance capabilities of organisms. However, the concept of heat and cold tolerance itself did not seem applicable to buildings as heat and cold tolerance in organisms would be unique to species while the comfort range in a building is not affected by the type of building. Envisaging buildings and species, occupants would normally function well only if the internal temperature is within their comfort range, a fact irrespective of the building type, size or age. In biology, research shows that in all species, small individuals have higher heat tolerance compared to large individuals and are thus capable of surviving under higher temperatures (Peck et al. 2009). However, the heat and cold tolerance of occupants can change slightly due to the acclimation of the human body to particular climates. For example, people who have been born and/ or have lived in cold climates for most of their life might have evolved more endurance to low temperatures and thus would feel comfortable in lower temperatures compared to people native to tropical environments. Athletes can also acclimatise to running in a hotter climate than they are used to and this process can take from 7–14 days depending on the regime used to become familiar with heat stress (Garrett et al. 2011).

The concepts of eurythermy and stenothermy also fell short of suggesting a classification scheme for biological thermal regulatory strategies. As discussed earlier, eurythermy and stenothermy describe the range of temperatures an organism can stand in a particular environment. The condition in which a building experiences a wide range of internal temperatures and still remains habitable could be seen as analogous to eurythermy. For example, Ooka (2002) monitored a traditional Japanese house in the Hokuriku area, where the temperatures varied from hot in summer to cold in winter. However, it was the warm-blooded residents that were able to adapt to these cold temperatures in winter "by controlling the amount of clothing or taking heat from a fire pit or other traditional heater such as a brazier through radiative heat transfer", since when the fire pit was lit the internal temperature only rose 1°C (Ooka 2002). Conversely, stenothermy might be analogous to an ordinary office building that keeps the temperature within a narrow band using a HVAC system.

While thermoregulatory concepts in organisms and buildings are similar in several aspects, some cannot be seen as applicable to buildings. The concepts of acclimation and acclimatisation are examples of such themes as these have almost no parallel in building design. One potential connection is the problem of simulating low-energy buildings with high mass that is insulated from the external environment, such as the development of sustainable earth-sheltered houses at Hockerton in the UK (Hockerton Housing Project 2020). In order to simulate the monitored temperatures using EnergyPlus, it was necessary to run the simulation for a year to allow the mass to heat up and its temperature to stabilise (Mithraratne and Vale 2006). This could be viewed as equivalent to the building becoming acclimatised to its environment.

Despite the fact that thermoregulatory mechanisms in nature evolve, buildings do not of themselves evolve over their lifetimes nor transfer any mutations to the next generations, although accumulation of knowledge by designers producing successive sustainable building designs could be seen as an imperfect evolutionary analogy. Additionally, while organisms migrate to more favourable climates as part of their phenotypic acclimation process, buildings do not tend to move in a similar way. Portable structures such as tents have been customised to adapt to different climates, with the black tent found in both hot dry climates as well as the cold climate of Tibet, where it is made of yak rather than goat hair (Manderscheid 2001). The tent could be seen as weak analogy, however, as portability in these cases is more to do with the search for food for domesticated animals than migration as part of thermal adaptation.

In nature, thermal adaptation happens through behavioural strategies and physiological changes in the bodies of organisms. While species with minimum capability for adaptation die and face extinction, those with maximum thermal tolerance become the best fit and thus prevail through natural selection. Even though fatality and extinction seem irrelevant in the context of building design, lessons learnt from successful sustainable strategies demonstrated by natural species can be used for designing the next energy-efficient building.

5.2.5 *Endothermy and ectothermy as a means of classification*

From the concepts discussed above, endothermy and ectothermy emerged as the most appropriate for categorising biological thermal adaptation strategies. The difference between endotherms and ectotherms is the location of the heat source on which they depend for thermal adaptation. Heat generation in the body of endotherms is parallel to heat generation in HVAC systems. Similar to organisms that use the thermogenesis effect of food to generate heat, HVAC systems consume electricity for heat production. Unlike endotherms, most ectotherms use behavioural adaptation strategies to regulate their body temperature.

Similarly, the voluntary (behavioural) and involuntary (autonomic or physiological) aspects of thermoregulation in organisms seemed useful as these made a sensible connection to the concepts of ectothermy and endothermy. In the latter, thermoregulation occurs within or through organ(s), tissue(s) and cell(s), and therefore is linked to the physiology of organisms. In fact, the majority of endotherms undergo physiological changes as part of thermoregulation. The living organism does not voluntarily respond to stimuli as physiological changes take place automatically. Therefore, autonomic or physiological thermal adaptation mechanisms are associated with physiological changes that are activated by external stimuli and are thus called 'active'. Conversely, ectotherms generally control their body temperature through their behaviour rather than physiological changes in their bodies. In this regard, thermal adaptation happens through external means and can thus be thought of as 'passive'.

Seeing buildings as living organisms, a sustainable building design strategy can be considered active when it causes changes in the building envelope or systems. For example, thermal regulatory strategies that occur through changes in circulatory systems by opening or closing valves or require energy for the mechanical control of an element like louvres for shading are active. Such changes require an energy input for them to happen. This is similar to physiological changes causing alterations in the circulatory systems of organisms. On the other hand, passive building design strategies are not associated with circulatory systems for heating and cooling. Essentially passive design strategies do not need additional energy for a thermal change to happen. The classic example is the south-facing glass façade of a passive solar house that does not change when the sun shines although the temperature inside the building will rise. These parallels suggest that in biology, thermal adaptation strategies can be called active when the heat is produced in the body of the organism and passive when the heat source is outside it. Given this, sustainable building design strategies can also be either passive or active depending on how a building adapts to its environment.

In view of this, the thermal adaptation of endotherms and ectotherms can also be explained by the ways in which acceptable internal temperatures are achieved. Heat conservation in ectotherms is primarily achieved through behavioural thermoregulatory strategies while endotherms use both autonomic and behavioural strategies for thermoregulation. For example, regional heterothermy in endotherms reduces heat loss from the peripheral tissues while thermoregulatory mechanisms for enhancing heat loss concern water evaporation from a wet surface. Examples of the latter are evaporation from the skin surface being wetted by sweating or the spread of saliva droplets (Needham et al. 1974). The cost of this evaporative cooling is that endotherms need to replace the lost water to stop the body from dehydrating. This becomes crucial for small endotherms using such a strategy (Albright et al. 2017).

Overall, the physiological thermal adaptation mechanisms that endotherms use to control heat exchange can be grouped under the following categories which are explained further in Chapter 6. These are:

1) circulatory mechanisms, that can be further classified into vasoconstriction and vasodilation, and counter current heat exchange,

2) insulation, such as fur, fat, or feathers, and

3) evaporative mechanisms, such as panting and sweating.

While behavioural thermoregulatory strategies are used by both endotherms and ectotherms, autonomic (physiological) thermoregulation is almost exclusive to endotherms. In addition, behavioural thermal adaptation strategies differ due to the medium surrounding the body of a particular animal. For example, the heat transfer coefficient of water is higher than that of air, which leads to rapid heat transfer in marine animals. Heat transfer is comparatively slower in land-based endotherms due to the somewhat lower thermal conductivity of the air. Therefore, in terms of behavioural thermoregulatory strategies, terrestrial animals need to exploit their environment in various ways to adjust the heat transfer. Davenport (2012) states that the behavioural thermal adaptation strategies animals use to decrease and increase heat gain can be categorised into basking, posture, orientation and locomotion, while clustering and huddling are used to produce heat. Behaviours such as seeking shade, migration and burrowing are used to prevent thermal stress while increased heat loss is behaviourally achieved by evaporative cooling (Davenport 2012).

The discussion suggested the concepts of endothermy and ectothermy were the most appropriate for a classification scheme for categorising the thermal adaptation strategies used by plants and organisms. This approach could be linked to the concepts of active and passive thermoregulation strategies, which were in turn derived from the definitions of voluntary and involuntary thermal regulation. The passive and active aspects of thermal adaptation could be applied to both living organisms and buildings and were thus used as means of dividing the thermal adaptation strategies produced in step 2 (5.2.2 and Figure 5-1) into active and passive groups. The details of how this was done are in Chapter 6.

References

Adey, W. H. and Loveland, K. (2011). *Dynamic Aquaria: Building Living Ecosystems.* Elsevier.

Airwolfhound. (2015). Fox - British Wildlife Centre (17429406401). Retrieved from https://commons.wikimedia.org/wiki/File:Fox_-_British_Wildlife_Centre_(17429406401).jpg, Creative Commons Attribution-Share Alike 2.0 Generic license: https://en.wikipedia.org/wiki/Creative_Commons.

Albright, T. P., Mutiibwa, D., Gerson, A. R., Smith, E. K., Talbot, W. A., O'Neill, J. J., McKechnief, A. E. and Wolf, B. O. (2017). Mapping evaporative water loss in desert passerines reveals an expanding threat of lethal dehydration. *National Academy of Sciences, 114*(9), 2283–2288. doi:10.1073/pnas.1613625114.

Angilletta Jr, M. J. and Angilletta, M. J. (2009). *Thermal Adaptation: A Theoretical and Empirical Synthesis.* New York, NY: Oxford University Press.

Bakken, G. S. (1976). A heat transfer analysis of animals: unifying concepts and the application of metabolism chamber data to field ecology. *Journal of Theoretical Biology, 60*(2), 337–384.

Beaufort, J. (n.d.). Grizzly Bear. Retrieved from https://commons.wikimedia.org/wiki/File:GrizzlyBearJeanBeaufort.jpg, Creative Commons CC0 1.0 Universal Public Domain Dedication: https://creativecommons.org/publicdomain/zero/1.0/deed.en.

Bowler, K. (2018). Heat death in poikilotherms: Is there a common cause? *Journal of Thermal Biology, 76*, 77–79.

Brain, C. and Mitchell, D. (1999). Body temperature changes in free-ranging baboons (Papio hamadryas ursinus) in the Namib Desert, Namibia. *International Journal of Primatology, 20*(4), 585–598. doi:Doi 10.1023/A:1020394824547.

Brakefield, P. M. (1990). A decline of melanism in the peppered moth Biston betularia in The Netherlands. *Biological Journal of the Linnean Society, 39*(4), 327–334.

Brown, R. T. and Baust, J. G. (1980). Time course of peripheral heterothermy in a homeotherm. *American Journal of Physiology-Regulatory, Integrative and Comparative Physiology, 239*(1), R126–R129.

Cabanac, M., Hildebrandt, G., Massonnet, B. and Strempel, H. (1976). A study of the nycthemeral cycle of behavioural temperature regulation in man. *The Journal of Physiology, 257*(2), 275–291.

Cheung, C. K., Fuller, R. J. and Luther, M. B. (2005). Energy-efficient envelope design for high-rise apartments. *Energy and Buildings, 37*(1), 37–48. doi:10.1016/j.enbuild.2004.05.002.

Clarke, A. (2017). *Principles of Thermal Ecology: Temperature, Energy and Life.* Oxford University Press.

Cossins, A. (2012). *Temperature Biology of Animals.* Springer Science & Business Media.

Davenport, J. (2012). *Environmental Stress and Behavioural Adaptation.* Sydney, Australia: Springer.

Davidvraju. (2013). Common Kukri Snake (Oligodon arnensis) (3). Retrieved from https://commons.wikimedia.org/wiki/File:Common_Kukri_Snake(Oligodon_arnensis)_(3).jpg, Creative Commons Attribution-Share Alike 4.0 International license: https://en.wikipedia.org/wiki/Creative_Commons.

Delmar Cerutti, R., Rizzo, M., Alberghina, D., Cristina Scaglione, M. and Piccione, G. (2019). Locomotor activity patterns of domestic cat (Felis silvestris catus) modulated by different light/dark cycles. *Biological Rhythm Research, 50*(6), 838–844.

Dupont, B. (2009). Granulated Sea Star (Choriaster granulatus) (8455465013). Retrieved from https://commons.wikimedia.org/wiki/File:Granulated_Sea_Star_(Choriaster_granulatus)_(8455465013).jpg, Creative Commons Attribution-Share Alike 2.0 Generic license: https://en.wikipedia.org/wiki/Creative_Commons.

Eagan, C. (1963). *Introduction and Terminology*. Paper presented at the Federation Proceedings.

Fernández, M. L., Rubio, R. and González, S. M. (2013). *Architectural Envelopes that Interact with their Environment*. Paper presented at the International Conference on New Concepts in Smart Cities: Fostering Public and Private Alliances (SmartMILE), Gijon, Spain.

Gamage, A. and Hyde, R. (2011). *Can Biomimicry, as an Approach, Enhance Ecologically Sustainable Design (ESD)?* Paper presented at the 45th Annual Conference of the Architectural Science Association: From Principles to Practice in Architectural Science, Sydney, Australia.

Garrett, A. T., Rehrer, N. J. and Patterson, M. J. (2011). Induction and decay of short-term heat acclimation in moderately and highly trained athletes. *Sports Medicine, 41*(9), 757–771.

GhaffarianHoseini, A. (2012). Ecologically sustainable design (ESD): theories, implementations and challenges towards intelligent building design development. *Intelligent Buildings International, 4*(1), 34–48.

Grimpo, K., Legler, K., Heldmaier, G. and Exner, C. (2013). That's hot: golden spiny mice display torpor even at high ambient temperatures. *Journal of Comparative Physiology B, 183*(4), 567–581.

Gzen92. (2019). Panda géant (Ailuropoda melanoleuca). Retrieved from https://commons. wikimedia.org/wiki/File:Panda_g%C3%A9ant_(Ailuropoda_melanoleuca)_(2). jpg, Creative Commons Attribution-Share Alike 4.0 International license: https:// en.wikipedia.org/wiki/Creative_Commons.

Halle, S. and Stenseth, N. C. (2012). *Activity Patterns in Small Mammals: An Ecological Approach* (Vol. 141): Springer.

Hensel, H., Brück, K. and Raths, P. (1973). Homeothermic organisms. *Temperature and Life*, 503–761.

Hockerton Housing Project. (2020). Celebrating 20 years of sustainable development at Hockerton Housing Project. Retrieved from https://www.hockertonhousingproject. org.uk/2018/09/celebrating-20-years-sustainable-development-hockerton-housing-project/.

Holmes, M. J. and Hacker, J. N. (2007). Climate change, thermal comfort and energy: Meeting the design challenges of the 21st century. *Energy and Buildings, 39*(7), 802–814. doi:10.1016/j.enbuild.2007.02.009.

Huey, R. B., Kearney, M. R., Krockenberger, A., Holtum, J. A., Jess, M. and Williams, S. E. (2012). Predicting organismal vulnerability to climate warming: roles of behaviour, physiology and adaptation. *Philosophical Transactions of the Royal Society B, Biological Science, 367*(1596), 1665–1679. doi:10.1098/rstb.2012.0005.

IUPS Thermal Commission, C. f. T. P. o. t. I. U. o. P. S. (1987). Glossary of terms for thermal physiology. *Pflügers Archiv-European Journal of Physiology, 410*, 567–587.

Ivanov, K. (2006). The development of the concepts of homeothermy and thermoregulation. *Journal of Thermal Biology, 31*(1-2), 24–29.

Jabareen, Y. (2008). A new conceptual framework for sustainable development. *Environment, Development and Sustainability, 10*(2), 179–192.

Jabareen, Y. (2013). Planning the resilient city: Concepts and strategies for coping with climate change and environmental risk. *Cities, 31*, 220–229.

Johansen, K. (1969). Adaptive responses to cold in arterial smooth muscle from heterothermic tissues of marine mammals. *Nature, 223*(5208), 866–867.

Karban, R. (2008). Plant behaviour and communication. *Ecology Letters, 11*(7), 727–739.

Katz, S. L. (2002). Design of heterothermic muscle in fish. *Journal of Experimental Biology, 205*(15), 2251–2266.

Kay, J. J. (2003). On complexity theory, exergy, and industrial ecology: some implications for construction ecology. pp. 96–131. *In*: C. J. Kibert, J. Sendzimir and G. B. Guy (eds.). *Construction Ecology: Nature as a Basis for Green Buildings*. New York, NY: Taylor & Francis.

Logan, C. A. and Buckley, B. A. (2015). Transcriptomic responses to environmental temperature in eurythermal and stenothermal fishes. *Journal of Experimental Biology, 218*(12), 1915–1924.

Loonen, R. C. G. M., Favoino, F., Hensen, J. L. M. and Overend, M. (2017). Review of current status, requirements and opportunities for building performance simulation of adaptive facades. *Journal of Building Performance Simulation, 10*(2), 205–223. doi:10.1080/1940149 3.2016.1152303.

Lovegrove, B. G. (2000). Daily heterothermy in mammals: coping with unpredictable environments. pp. 29–40. In: *Life in the Cold*. Springer.

Majerus, M. E. (2009). Industrial melanism in the peppered moth, Biston betularia: an excellent teaching example of Darwinian evolution in action. *Evolution: Education and Outreach, 2*(1), 63–74.

Maloney, S. K., Moss, G. and Mitchell, D. (2005). Orientation to solar radiation in black wildebeest (Connochaetes gnou). *Journal of Comparative Physiology A, 191*(11), 1065–1077. doi:10.1007/s00359-005-0031-3.

Manderscheid, A. (2001). The black tent in its easternmost distribution: The case of the Tibetan Plateau. *Mountain Research and Development, 21*(2), 154–160.

Martin, P., Rancon, J., Segura, G., Laffont, J., Boeuf, G. and Dufour, S. (2012). Experimental study of the influence of photoperiod and temperature on the swimming behaviour of hatchery-reared Atlantic salmon (Salmo salar L.) smolts. *Aquaculture, 362*, 200–208.

Mitchell, D., Snelling, E. P., Hetem, R. S., Maloney, S. K., Strauss, W. M. and Fuller, A. (2018). Revisiting concepts of thermal physiology: Predicting responses of mammals to climate change. *Journal of Animal Ecology, 87*(4), 956–973. doi:10.1111/1365-2656.12818.

Mithraratne, K. and Vale, B. (2006). Modelling of thermal characteristics of insulated mass in zero-heating passive solar houses: Part 2—simulation results. *Architectural Science Review, 49*(3), 221–228.

Mole, M. A., Rodrigues DÁraujo, S., Van Aarde, R. J., Mitchell, D. and Fuller, A. (2016). Coping with heat: behavioural and physiological responses of savanna elephants in their natural habitat. *Conservation Physiology, 4*(1), 1–11.

Moore, J. A. (1940). Stenothermy and eurythermy of animals in relation to habitat. *The American Naturalist, 74*(751), 188–192.

Muller, F. (2000). *Handbook of Ecosystem Theories and Management*. CRC Press.

National Geographic. (2020). Adaptation. Retrieved 22 October 2020 https://www.nationalgeographic.org/encyclopedia/adaptation/.

Needham, A. D., Dawson, T. J. and Hales, J. R. (1974). Forelimb blood flow and saliva spreading in the thermoregulation of the red kangaroo, Megaleia rufa. *Comparative Biochemistry and Physiology, 49*(3A), 555–565.

Nord, A. and Folkow, L. P. (2018). Seasonal variation in the thermal responses to changing environmental temperature in the world's northernmost land bird. *Journal of Experimental Biology, 221*(1).

Ooka, R. (2002). Field study on sustainable indoor climate design of a Japanese traditional folk house in cold climate area. *Building and Environment, 37*(3), 319–329.

Peck, L. S., Clark, M. S., Morley, S. A., Massey, A. and Rossetti, H. (2009). Animal temperature limits and ecological relevance: effects of size, activity and rates of change. *Functional Ecology, 23*(2), 248–256.

Pincebourde, S., Murdock, C. C., Vickers, M. and Sears, M. W. (2016). Fine-scale microclimatic variation can shape the responses of organisms to global change in both natural and urban environments. *Integrative and Comparative Biology, 56*(1), 45–61. doi:10.1093/icb/icw016.

Pixabay. (n.d-a). Black Bear. Retrieved from https://pixabay.com/photos/black-bear-bear-louisiana-937037/, Pixabay License: https://pixabay.com/service/license/.

Pixabay. (n.d-b). Map of the World. Retrieved from https://pixabay.com/illustrations/map-of-the-world-earth-world-map-862718/, Pixabay License: https://pixabay.com/service/license/.

Pörtner, H. O. and Gutt, J. (2016). Impacts of climate variability and change on (marine) animals: physiological underpinnings and evolutionary consequences. *Integrative and Comparative Biology, 56*(1), 31–44.

Riek, A. and Geiser, F. (2014). Heterothermy in pouched mammals—a review. *Journal of Zoology, 292*(2), 74–85.

Rummel, A. D., Swartz, S. M. and Marsh, R. L. (2019). Warm bodies, cool wings: regional heterothermy in flying bats. *Biology Letters, 15*(9), 20190530.

Segundo-Ortin, M. and Calvo, P. (2019). Are plants cognitive? A reply to Adams. *Studies in History and Philosophy of Science Part A, 73*, 64–71.

Smith, R. S., Guyomarc'h, S., Mandel, T., Reinhardt, D., Kuhlemeier, C. and Prusinkiewicz, P. (2006). A plausible model of phyllotaxis. *Proceedings of the National Academy of Sciences, 103*(5), 1301–1306.

Stawski, C., Willis, C. and Geiser, F. (2014). The importance of temporal heterothermy in bats. *Journal of Zoology, 292*(2), 86–100.

Taiz, L., Alkon, D., Draguhn, A., Murphy, A., Blatt, M., Hawes, C., Thiel, G. and Robinson, D. G. (2019). Plants neither possess nor require consciousness. *Trends in Plant Science, 24*(8), 677–687.

Taylor, N. A. (2006). Ethnic differences in thermoregulation: genotypic versus phenotypic heat adaptation. *Journal of Thermal Biology, 31*(1-2), 90–104.

Todd, N. J. and Todd, J. (1994). *From Eco-Cities to Living Machines: Principles of Ecological Design*. North Atlantic Books.

Trier, T. M. and Mattson, W. J. (2003). Diet-induced thermogenesis in insects: a developing concept in nutritional ecology. *Environmental Entomology, 32*(1), 1–8.

Tzschentke, B. and Rumpf, M. (2011). Embryonic development of endothermy. *Respiratory Physiology & Neurobiology, 178*(1), 97–107.

Valladares, F. and Brites, D. (2004). Leaf phyllotaxis: Does it really affect light capture? *Plant Ecology, 174*(1), 11–17.

Van der Ryn, S. and Cowan, S. (2013). *Ecological Design*. Island press.

Varner, J. and Dearing, M. D. (2014). The importance of biologically relevant microclimates in habitat suitability assessments. *PLoS One, 9*(8), e104648. doi:10.1371/journal.pone.0104648.

VJAnderson. (2019). Douglas Squirrel DSC3742vvc. Retrieved from https://commons.wikimedia.org/wiki/File:Douglas_Squirrel_DSC3742vvc.jpg, Creative Commons Attribution-Share Alike 4.0 International license: https://en.wikipedia.org/wiki/Creative_Commons.

Waheed, B., Khan, F. and Veitch, B. (2009). Linkage-based frameworks for sustainability assessment: making a case for driving force-pressure-state-exposure-effect-action (DPSEEA) frameworks. *Sustainability, 1*(3), 441–463.

Waltner-Toews, D., Kay, J., Kay, J. J. and Lister, N.-M. E. (2008). *The Ecosystem Approach: Complexity, Uncertainty, and Managing for Sustainability*. Columbia University Press.

Waterhouse, J. (2013). Homeostatic control mechanisms. *Anaesthesia & Intensive Care Medicine, 14*(7), 291–295.

Willmer, P., Stone, G. and Johnston, I. (2009). *Environmental Physiology of Animals*. John Wiley & Sons.

Woods, S. C. and Ramsay, D. S. (2007). Homeostasis: beyond Curt Richter. *Appetite, 49*(2), 388–398.

Chapter 6
Thermoregulation in Nature

6.1 Introduction

This chapter uses the active and passive aspects of thermal adaptation as a systematic means of categorising the thermal adaptation strategies used by living organisms. The first two sections (6.2 and 6.3) explain how animals adapt to their environment thermally. This is then followed by introducing the thermoregulatory mechanisms used by plants (6.4).

Grouping the thermoregulatory strategies for animals is different from plants, as it is unclear how to distinguish active and passive strategies in the latter, while this is easy to determine for animals. The determination of whether a thermoregulatory mechanism is either active or passive is in line with the definition provided in chapter 5, where a thermoregulatory strategy is active if the heat is produced in the body of organism. While some plants do use metabolic heat generation to thermoregulate, the literature on whether plants could be considered to be endotherms is diverse and scattered. As suggested by Michaletz et al. (2015), for plants, "…the implications of thermoregulation have been difficult to implement. Many studies in physiology, ecology, and climate science still regard plants as poikilotherms—with temperatures that are determined solely by the environment." This is discussed further in section 6.4.

Sections 6.2 and 6.3 cover how heat gains and losses are controlled in the bodies of animals through both passive and active means of thermal adaptation. Since heat is generated through metabolic processes in the bodies of animals and some insects, the thermal regulatory strategies these organisms use to generate heat are also described. Further, for both heat loss and heat gain three subcategories are introduced to cover the strategies that animals use to control each of these through decreasing, increasing and avoiding heat transfer.

Section 6.1 introduces the passive methods of thermal adaptation used by animals. The hierarchical categorisation of these solutions is shown in Figure 6-1. Column 'A' presents the main action for heat control called the 'parent action' with the three categories being (1) generating heat, (2) controlling heat gain and (3) controlling heat loss. In column 'B', controlling heat gain and controlling heat loss are further broken down into the sub-actions of decreasing, increasing and preventing heat gain,

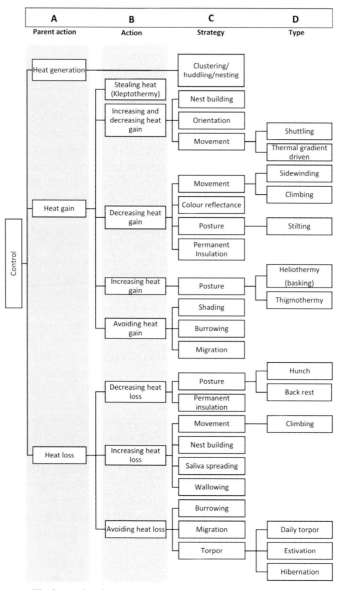

Figure 6-1. The hierarchical categorisation of passive thermal adaptation solutions.

and decreasing, increasing and avoiding heat loss. Column 'C' introduces the biological solutions for some of these, showing how thermal adaptation takes place in different ways. These are called the 'types' of biological solution and are shown in column 'D'. Some strategies have sub-types which are shown in column 'E'.

6.2 Controlling heat: passive methods of thermal adaptation in animals

6.2.1 Generating heat

Huddling, clustering and nesting or nest sharing are used to generate heat. Huddling is the main term used to describe social thermoregulation. However, other terms have been used to explain this behaviour in different contexts. The choice of which term to use seems to be related to the characteristics of the group in question, such as the density, the number of species involved in the aggregation and the duration of the aggregation (Gilbert et al. 2010). Some of these terms are clustering, clumping, grouping, communal nesting, nest sharing, communal roosting and crèching.

Huddling is prevalent among endotherms (Brown and Foster 1992, Arends et al. 1995). In fact, people often talk about huddling together for warmth. However, two factors that affect thermoregulatory heat generation are variations in the form of the species involved and the population of the cluster. For example, in small mammals, the proportion of the exposed body area to body volume impacts the efficiency of the huddle for heat generation (Canals et al. 1989). Male emperor penguins join a huddle in order to generate heat and around 5,000 penguins (approximately 10 per square metre) move slowly downwind and leeward. Once they stop, steam can be seen rising from the huddle (Thomas and Fogg 2008). Young animals often huddle for warmth; Shea (2019) describes how a grey wolf pack on Ellesmere Island in the Canadian Arctic that had recently fed lay down to sleep with the "pups piled together in a downy mess". Collective overwintering behaviours such as huddling can be seen in socialised species. For example, bees are social insects and will cluster together in the hive in winter to conserve heat (Seeley 2010). There are a variety of arrangements of bees in a cluster, performing together as a porous mobile medium between the core and the environment, to achieve thermoregulation. Ocko and Mahadevan (2014) studied the conductivity and permeability of bee clusters by measuring the packing fraction and ambient air temperature.

The effect of huddling has also been explored at the cellular level. In recent research, an experiment was conducted on voles to study the impacts of huddling behaviour in reducing the damaging effects of low ambient temperature on mammalian myocardium (Wang et al. 2020). Myocardium is the heart muscle made up of mitochondria, which are cell organelles that play a significant role in homeostasis in metabolic tissues. In voles, reduced temperature damages mitochondria and can lead to death. Research has shown that the mitochondria energy supply was higher when voles huddled as an adaptive behaviour (Wang et al. 2020).

In some mammals, huddling behaviour is used to reduce the cost of thermogenesis. For some species, during huddling there is a difference between the temperatures of the internal and surface tissues. For example, southern elephant seals use aggregation behaviour during the moult stage (Chaise et al. 2019). While they usually use the insulation capability of their fur as part of the control of heat exchange, they cannot use this when moulting. To huddle, a large population will gather in mud pools where the higher ground temperature and lower wind speed creates a favourable microclimate. The number of seals and hence, size of the huddle, also increases in lower temperatures.

Huddling seems to have been used in conjunction with other thermal adaptation strategies while the influential factors that lead to huddling have been less explored. For example, research shows that huddling facilitates spontaneous torpor in the large Japanese field mouse but it is not clear what mechanisms induce such huddling (Eto et al. 2014). More research needs to be conducted to explore the complex relationship of interacting thermal adaptation strategies and the parameters that activate them.

Species that have not developed fur for insulation employ nest sharing as a strategy for surviving cold climates (Crawford 2013), and this is prevalent amongst small mammals like lemmings and voles. Again, the decrease in exposed bodily surface reduces heat loss (Vaughan et al. 2013), so this behaviour saves energy.

6.2.2 Controlling heat gain

Heat from the environment such as absorbing solar energy can be useful for endotherms as less energy is expended on maintaining internal heat. However, too much heat can lead to heat stress, so other thermoregulatory strategies are often employed to control heat gain.

Stealing heat (Kleptothermy): Brischoux et al. (2009) identified kleptothermy as an unusual thermoregulatory strategy used by both endotherms and ectotherms. Kleptothermy is a strategy where an animal steals heat from other animals to keep its body warm. This can be across species; for example, some species of snakes share their burrows with seabirds in order to raise their body temperature. The heat is transferred either through direct bodily contact or via the air. The latter is a convective heat transfer method in which the surrounding air is heated by the respiration and metabolism of the donor organism (the seabird) occupying the same space. Obviously, moving away from the source of heat ensures control of the heat gain.

While some species might benefit from kleptothermy, others can be negatively affected. For example, in the observation of two species of mussels known as indigenous *Perna perna* and the invasive *Mytilus galloprovincialis*, it appeared that stealing heat was beneficial to the latter while it was a life-threatening situation for the former in the face of heat stresses (Seuront et al. 2018). These two species coexist in large clusters on the south coast of South Africa but have different patterns of respiration. *Perna perna* frequently opens its shell when moving to the surface water. This behaviour is known as gaping and enables the species to maintain respiration but simultaneously increases water evaporation which might lead to desiccation. Conversely, *Mytilus galloprovincialis* keep their valves closed when exposed to air. While the temperature of a solitary mussel does not decrease significantly by the evaporation and expulsion of water from this gaping behaviour, a cluster of *Perna perna* can lose a great deal of cumulative heat to reduce their temperature and thus can survive when the environmental temperature rises. Where *Perna perna* aggregate in groups but are covered by *Mytilus galloprovincialis,* they lose this opportunity of decreasing their body temperature through evaporation (Nicastro et al. 2012).

Increasing and decreasing heat gain: The thermoregulation strategies outlined below have the dual effects of decreasing and increasing heat gain. To avoid repetition, they have been grouped and introduced as one.

*Nest building: the parameters affecting nest thermoregulation are population size, moisture and the thermal conductivity of the nest materials. The mounded nest building of ants exemplifies the latter as "...the unique thermal and insulation properties of the nest material help to maintain stable conditions" (Kadochová and Frouz 2014). Other nests show different patterns of heat control (Section 6.2.1).

The orientation of termite nests and those of other ant species is an significant factor in moderating nest temperature (Lane and Skaer 1980). Thermoregulation is even affected by the shape of the nest. Fire ants demonstrate this when they build an oval-shaped nest with the majority of the long axes extending from south to north. The alignment of the external slope of the surface of the nest to the sun's rays can increase solar heat absorption. The position and orientation of the nest entrance can also affect the range of temperatures within. It has been hypothesised that "yearly mean temperature influences entrance orientation in birds, nests at higher latitudes are expected to face the equator in order to take advantage of warmer temperatures" (Schaaf et al. 2020).

Looking at these examples, it is obvious that nest building is effective in controlling heat through a combination of orientation, shape and the material properties of the nest. Orientation is introduced as a separate thermal adaptation strategy below.

Orientation: While obvious, it is important to include the fact animals will change their orientation to control heat gain as they bask in the sun. As they change their position, they alter the area of exposed body surface and therefore regulate temperature (Kevan and Shorthouse 1970, Penacchio et al. 2015). However, recent research shows that for some vertebrates like lizards, the body alignment cannot be explained by the position of the sun but rather as an alignment to the earth's magnetic fields (Diego-Rasilla et al. 2017).

Movement: Changing orientation involves movement and these movements can be further categorised as:

- *Shuttling*: This is a rapid movement. The desert lizard *Uromastyx aegyptia microlepis* regulates its body temperature by rapidly moving between sun and shade, and thus between hot and cold microenvironments (Withers and Campbell 1985).

- *Moving along a thermal gradient*: The watery habitat of semi-aquatic amphibians will govern their behaviour. Because they are restricted by their aquatic habitats, they do not have prominent thermoregulatory behaviours. To control heat gain, they simply move to a warm or cool area of their watery habitat. However, as the supply of oxygen may not always be as plentiful in parts of the habitat, a balancing act is required for those species that are partly dependent on cutaneous respiration. These behaviours are known as predictive thermoregulation (Davenport 2012).

Decreasing heat gain: Some strategies are only used for losing heat and these are set out below.

Movement: Movement is also important when an animal is trying to reduce heat gain.

- *Sidewinding*: Rather than sliding in a straight line, some snakes will move from side to side as a way of enabling them to move rapidly over hot surfaces. This type of movement minimises both the time each part of the skin is in contact with the hot surface and the area of surface contact. This is because a "...sidewinding snake simultaneously bends and lifts its body while maintaining two regions of static contact between its belly and the ground" (Secor et al. 1992).

- *Climbing*: Several lizards also climb to avoid heat uptake and also increase heat loss by evaporative cooling. Snails have also been reported as climbing to reduce heat uptake by increasing the distance from the heat source (Dittbrenner et al. 2008). Another example is the pipevine swallowtail, a species of caterpillar that lives and feeds on low-growing plants. When the temperature increases, these caterpillars climb to a cooler location above their host vegetation to decrease heat gain. This is a thermal refuge-seeking behaviour that is known to be effective in cooling (Nielsen and Papaj 2015). While this could be a useful strategy for small caterpillars, it might not be beneficial to large species. Like other thermal adaptation mechanisms, climbing involves energy so there is an energy trade-off between the rate of energy expenditure and the cooling obtained.

Colour reflectance: Dark colours absorb more heat. Ghassemi Nejad et al. (2017) have shown that various skin or coat colours are more effective in absorbing or reflecting solar radiation. However, colour is not the only issue as the ability of the skin to absorb near-infra red radiation also plays a part in thermoregulation (Stuart-Fox et al. 2017). The same authors also note that "The relationship between 'colour' and thermoregulation is more complex in endotherms than ectotherms because of the presence of an insulating layer over most of the body." This again suggests that many aspects may have to be considered together when it comes to thermoregulation.

Posture: Some animals such as sheep will change to a lying or standing posture to control heat transfer (Mannuthy 2017). Similarly, a type of desert lizard does the same through the way it postures.

- *Stilting*: Stilting is an extra ability that some species, such as locusts, have to survive high temperatures. Standing reduces the area of contact between locusts and the ground, minimising heat transfer. In even higher temperatures, locusts climb local vegetation, increasing distance between the hot surface and themselves (Seeley 2010). Different species of insects use stilting behaviour, including extending some or all of their legs. Such behaviour has also been observed in animals (Dean and Milton 1999). For example, some geckos seek cooling by creating a space for air movement between their bodies and a hot rock surface by lifting their bodies away from the rock. This also reduces heat transfer by surface contact.

Permanent insulation: The insulation properties in the coats of birds and mammals affect solar energy absorption. For example, the complex coat structure of rock squirrels when exposed to intense solar radiation is optimised to control heat gain. This heat gain optimisation is associated with the specific ratio of their inner to outer coat (Walsberg 1988). The quantified ratio is close to the predicted value which in theory is needed to minimise solar heat gain.

Overall, solar heat gain in mammals and birds is influenced by a series of variables which together adjust the heat absorption. Heat gain in these species is a "function of coat properties, including structure, insulation, short-wave reflectivity, hair or feather optics and skin colour" (Walsberg 1988). This is a broad area to be investigated as "the functions of fur are considerable and diverse and evolutionary compromises associated with mammalian coats could be many" (Dawson and Maloney 2017). Such compromises are made to adjust the body temperature of a species through their aggregate effects on thermoregulation.

Increasing heat gain

Posture: Posture is also useful for increasing heat gain. Ectotherms use basking or heliothermy and thigmothermy to control the heat balance in their bodies.

- *Basking/heliothermy:* The IUPS Thermal Commission (1987) defines basking or heliothermy as "The regulation of the body temperature of an ectothermic animal by behavioural adjustments of its exposure to solar radiation". In basking, animals intelligently orientate their bodies in order to absorb maximum solar heat, or to reduce exposure after achieving an optimal body temperature (DeWitt et al. 1967). While heliothermy seems a preliminary thermal adaptation strategy, it has been seen as "an important aspect of the biology of an organism, necessitating the evolution of specialized morphology, physiology, ecology, and behaviour" (Griffing et al. 2020). However, this behavioural strategy may only be beneficial for a short time. Too much exposure to heat can also be a problem, especially for ectotherms that live in cold regions, where global warming is a huge threat. Coping with the effects of climate change requires physiological thermal adaptation mechanisms, which is a problem for these species as they mostly use behavioural thermoregulatory strategies, and this might lead to their extirpation by 2070 (Vicenzi et al. 2019).

  While most research has been focused on the behavioural adjustments of ectotherms, some research has been conducted on biophysical mechanisms, such as adjusting skin perfusion and coloration. Doing this can reduce the energy and time allocated to basking. Common lizards (*Zootoca vivipara*) living at high altitude will also alter their basking sites or their preferred temperature to reduce heat gain (Trochet et al. 2018).

- *Thigmothermy:* Thigmothermy is when animals, mainly nocturnal species that have limited access to sunlight, draw heat from something previously warmed by solar radiation. In a forest environment, some nychthemeral species practice thigmothermy. Although tree canopies block most direct sunlight, animals can make use of heat trapped within tree canopies (Garrick 2008).

Avoiding heat gain (in place)

Shading: Animals seek shelter to avoid overheating. In deserts, animals use the shade provided by rocky outcrops to avoid heat from the sun. Farm animals will also seek the shade of trees and farm buildings if the temperature increases (Ratnakaran et al. 2017).

Burrowing: some mammals and reptiles will use the more stable temperature of a burrow as a means of avoiding heat gain (Wilms et al. 2011). Burrows are thermally buffered microclimates which allow animals to survive (Milling et al. 2018) and are an important part of the acclimatisation of animals to their environment. The orientation and morphology of burrows affect their ventilation (Ganot et al. 2012). The temperature and humidity of underground microclimates is influenced by a number of factors including the dimensions, shape and depth of the burrow, vegetation cover and soil porosity. Soil properties such as colour and thermal conductivity influence the level and rate of heat absorption by solar radiation and hence, the temperature fluctuation cycle. Generally, the deeper a burrow is below ground level, the more consistent its temperature. The diameter, length and shape of burrows, as well as the difference in depth between interconnected tunnels, all impact ventilation, as a larger difference in depth between interconnected tunnels results in better ventilation (Burda et al. 2007).

There are three ways in which air advection can be achieved in burrows (in all of which a temperature gradient induces airflow): a pressure gradient along a burrow induced by wind, a piston-effect movement by the burrow's inhabitants and thermal conductive venting (Ganot et al. 2012). A comprehensive study by Knaust (2012) categorized the morphology of burrows (Figure 6-2), suggesting the complex geometric patterns could inspire innovative ventilation strategies for buildings.

While burrowing is effective for avoiding both heat gain and heat loss, it involves a significant degree of effort. Thus, animals will use this strategy only if the energy expenditure of burrowing is less than that of moving about to keep warm (Wu et al. 2015). The same authors point out that energy cost is an important factor for organisms as "terrestrial locomotion, such as running, is more energetically expensive than flying or swimming for an animal of similar mass, but considerably less costly than travelling through a dense, cohesive medium such as soil". Research conducted to observe the burrowing strategies of lizards showed a relationship between the body shape of an organism and climate (Grizante et al. 2012), while the morphology of an organism also influences the size of the burrow.

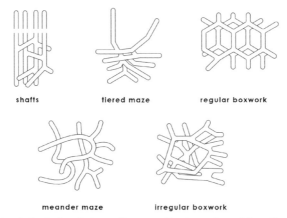

shafts   tiered maze   regular boxwork

meander maze   irregular boxwork

Figure 6-2. Morphological variations of burrows (Author, adapted from Frey et al. (1978)).

Migration: Some animals migrate as a thermoregulatory behaviour. The Atlantic Bluefin Tuna, for example, show migratory behaviour linked to finding a water temperature suitable for breeding (Block et al. 2001). Blue sharks also migrate; "...the overwintering grounds of immature blue sharks from the northern sector of the northwest Atlantic apparently lie in the warm waters of the Gulf Stream and in the central North Atlantic as far south as the Sargasso Sea" (Campana et al. 2011). Likewise, during hot periods, some bird species fly to cooler microclimates (Newton 2010).

Heat storage

Gigantothermy: Gigantothermy is also known as ectothermic homeothermy or inertial homeothermy. It is a strategy found in palaeontology whereby large ectothermic animals could more easily maintain a constant and relatively high body temperature compared to smaller animals, even at very cold temperatures. For giant ectothermic marine animals, the temperature of the surrounding water changes in a vertical gradient and gets extremely cold in the deeper parts of the ocean. While some gigantic animals swim back and forth to catch their prey, their body temperature does not change even in the deepest part of the ocean as gigantothermy allows heat conservation in the muscles of animals, thus enabling them to survive in super-cold temperatures. Although this is a hypothetical thermoregulatory process in large dinosaurs, it has been suggested that it is also used by leatherback turtles (Paladino et al. 1990) and endotherms such as Asian elephants (Rowe et al. 2013). It has been assumed that for endotherms using gigantothermy in hot climates, the metabolic heat generation reduces as their size increases. This means as adults, they do not need as much food as their peripheral tissues store heat to keep them safe in both tropical and cold climates. However, overheating can lead to the death of the animal (Currie 2017). For example, it has been hypothesised that for dinosaurs, the excess heat was dissipated from the neck and the tail (Paladino et al. 1990).

There seems to be a difference between gigantothermy and the strategy of 'insulation', as in the former, heat is stored in the muscles and needs to be released from other organs. On the other hand, in the latter, there is a layer that slows down heat loss from the warm body, or heat gain to the body. While insulation is more associated with decreasing heat gain and heat loss, gigantothermy deals with heat storage.

6.2.3 Controlling heat loss

Decreasing heat loss

Posture: Animals living in environments with fluctuating temperatures use their posture to adjust their body temperature. While adopting different postures has thermal benefits for animals, there are always other costs involved and it is essential to study the trade-offs associated with thermoregulation (Yorzinski et al. 2018):

- *Hunching*: Animals decrease heat loss by changing their body posture. Piglets, for example, crouch to keep warm (Monteith and Mount 1973). Many mammals hunch in a ball-like posture to reduce the surface to volume ratio of their bodies (Sultan 2015). Hunching as a thermoregulatory behaviour is usually observed when ambient air temperatures fall, and is practised by a number of organisms including monkeys, lemurs, rodents and seals (Terrien et al. 2011).

- *Back rest (head-tuck and leg-tuck)*: These postures are common among birds as they either withdraw their feet into plumage or tuck their head and neck under their wings to reduce heat loss. However, heat loss is not the same for the different regions of their bodies. For example, in peafowls, most heat is lost through the head while legs are the second significant contributors to heat loss. The heat loss from the neck, beak and body do not differ substantially (Yorzinski et al. 2018). The back rest strategy mostly occurs in cooler temperatures but is less common among birds with small bills and highly prevalent in wild populations such as shorebird species (Ryeland et al. 2017).

Permanent insulation: Mammals insulate their bodies to decrease heat loss with fur or fat. For some mammals, fur also acts as a barrier which reduces heat gain from the environment. Most mammals have hair or fur which traps a layer of air close to their skin, functioning as an insulating layer by reducing heat transfer. Marine mammals such as whales have an additional thick layer of fat called blubber. In very cold climates, the fat layer is important but for the Antarctic Emperor Penguin, feathers are also vital. "Insulation is provided by double-layered, high density feathers and a 2–3-cm-thick layer of subcutaneous fat. The thermal conductivity of fat is a quarter of that of water but the feathers, with the air layer which they entrap, provide more than 80% of the thermal insulation" (Thomas and Fogg 2008).

Polar terrestrial invertebrates, plants and microbes are less sensitive to low temperatures than species living in non-polar latitudes. One of the most significant adaptation strategies in the polar climate is insulation. For example, despite its small size, the arctic fox possesses the most effective fur insulation (Barnes 2012). The larger wolf also has fur that has evolved to keep the animal warm in very low temperatures. The wolf coat has two layers of the longer guard hairs that protect from wind, rain and snow, and an under layer of thick, soft fur that traps air for insulation. Likewise, polar bear fat and fur have appropriate insulation capacities, which are approximately seven times more effective than the fat layer of humans and the insulation from their clothing. The high thermal insulation capability of polar bear hair is because of its unique microstructural pattern consisting of a hollow, air-filled core and aligned shell. The core is porous, but the aligned surrounding shell gives it sufficient mechanical support (Cui et al. 2018). Polar bear fur has specific properties of transparency to short-wave radiation and colourlessness, which together with a dark skin, enables polar bears to absorb heat (Thomas and Fogg 2008).

For some mammals, the crimping structure of hairs plays an important role in heat retention. For example, yak hairs have multiscale structures with the outer surface of the skin usually covered by coarse hairs that are surrounded by many fine hairs. The latter are long and have various types of spiral crimps. Depending on the types of crimp, the reflectance of infrared light varies in order to control the thermal insulation of the hair strands. Unlike polar bear hair, the cross-section of a yak hair is solid and instead the high thermal insulation is influenced by the thermal properties of the hair strands in their different layers. This means, the thermal insulation provided by yak hairs changes in a hierarchical order depending on the shape of the crimps, which also vary from the root of each hair strand to its head (Xiao et al. 2019).

Increasing heat loss

Movement: Some animals move to increase heat loss.

- *Climbing*: The koala mainly eats eucalyptus leaves, which are low in nutrients so the animal rests for much of the day. In very hot weather, the koala will climb high up in a tree and rest in its shade until the temperature falls (Lunney et al. 2012) although this could perhaps be better viewed as decreasing heat gain. However, as wind speed increases with distance from the ground, the koala may also benefit from increased heat loss from greater cooling breezes than at ground level.

Nest building: In termite mounds, the structure can allow both open and closed ventilation. Open ventilation occurs when a series of holes in the walls of a nest allow air to stream in and out (Korb and Linsenmair 2000). This is possible because holes are made at different heights, which are then exposed to different wind velocities, resulting in a temperature gradient. In contrast, closed ventilation systems use peripheral air channels or cavities, which extend from the bottom to the top of the nest, rather than holes.

Saliva spreading: The spreading of saliva to cool the surface of an organism by evaporation is usually a deliberate behavioural, thermoeffector action (Needham et al. 1974). This is sometimes inaccurately referred to as grooming.

Wallowing: According to the IUPS Thermal Commission (1987), wallowing is "The thermoregulatory increase in evaporative heat loss by spreading an aqueous fluid (e.g., water, mud, urine) on the body surface". Buffaloes, for example, spend more time wallowing when the sun reaches its maximum strength at midday compared to other times of day (Mannuthy 2017). Likewise, the desert iguana, *Dipsosaurus*, and the desert locust, *Schistocerca*, wriggle their abdomens in sand to keep cool.

Avoiding heat loss

Burrowing and Migration: Although a key strategy animals use for surviving cold temperatures is keeping a low-level metabolism (Barnes 2012), such as bears hibernating during winter, retreating to the more stable temperature of a burrow is also a means of avoiding unnecessary heat loss. For example, crabs living in the polar regions take advantage of the thermal stability of burrows that can be found in deep-water sediment matrix (Watson et al. 2018). Burrowing is a key thermal adaptation strategy for small mammals such as rabbits as they do not have enough fat to insulate their bodies, nor do they huddle or hibernate (Whittaker and Thomas 1983).

Many birds will overwinter in warmer climates and return to their cooler breeding grounds in the spring. For example, the Greenland barnacle goose (*Branta leucopsis*) "breeds in north-east Arctic Greenland and migrates to Ireland and Scotland for winter, staging in Iceland during spring and autumn" (Doyle et al. 2020).

Torpor: This occurs when an organism assumes a state of inactivity and is much less responsive to what is happening in the external environment, such as estivation during the summer or hibernation during the winter.

- *Estivation*: Estivation or Aestivation is a prolonged torpor which occurs when ectotherms experience a lack of energy specifically in summer. However, estivation is not restricted to ectotherms. The number of daily torpor events during the aestival period increases for large-bodied species when environmental conditions are more restrictive in terms of water availability and ambient temperature (Valera et al. 2011). Estivation in some species could be linked to their particular morphology. For example, the limbless morphology of snakes causes them to have a higher body temperature compared to animals which are able to use posturing to control heat gain. Thus, these animals show greater tendency towards estivation (Seigel et al. 1987).

- *Hibernation*: Hibernation is a multi-day torpor which heterothermic mammals use, presumably as a strategy to conserve energy. Heterothermic mammals can change between self-regulating their body temperature and allowing the external environment to influence it.

6.3 Controlling heat: active methods of thermal adaptation in animals

This section covers the active thermal adaptation strategies used by animals with the hierarchical categorisation of these shown in Figure 6-3. As discussed in Chapter 5, a thermal adaptation strategy can be called active when it is associated with a change taking place in the body of an organism. These changes are involuntarily and linked to the physiology of animals. These changes occur in the cells, tissues and organs of the body.

6.3.1 Generating heat

Although ectotherms generally change their body in response to their environment, survival in extremes of cold or heat may not be possible. This has been suggested as the reason for some ectotherms having mechanisms for adjusting their body temperature irrespective of the thermal condition of their environment (Davenport 2012). Examples include tuna fish, which have a counter-current blood circulatory system which results in an increase in the temperature of their bodily muscles (Legendre and Davesne 2020). Some species, such as turtles, large crocodilians and several species of flying insects, maintain their body temperature above the ambient temperature of their environment through muscular thermogenesis.

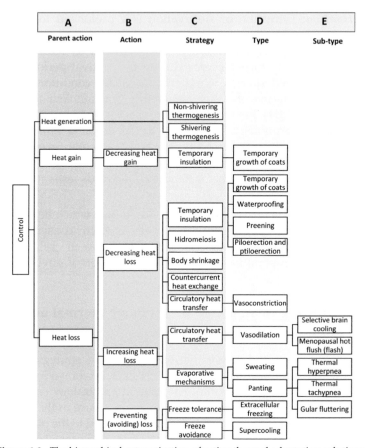

Figure 6-3. The hierarchical categorisation of active thermal adaptation solutions.

Generally, heat generation in animals can be categorised into two groups: non-shivering and shivering thermogenesis.

Thermogenesis, non-shivering: "[is] the increase in non-shivering thermogenesis in response to acute cold exposure. The principal effector organ is the brown adipose tissue which may adaptively increase its capacity for heat production in the course of acclimat(isat)ion and adaptation to cold stress" (IUPS Thermal Commission 1987). Brown adipose tissue, also known as brown fat, has a heat-generating capacity and occurs in mammals, particularly animals with small bodies and the new-born offspring of larger species such as humans.

Thermogenesis, shivering: "[is] an increase in the rate of heat production during cold exposure due to the increased contractile activity of skeletal muscles not involving voluntary movements and external work" (IUPS Thermal Commission 1987).

6.3.2 Controlling heat gain

There do not seem to be any active thermal adaptation strategies used by animals for controlling heat gain. Excess heat will have a physiological effect and will eventually lead to death. The mechanisms for losing heat, such as sweating, are discussed below under the heading 'Increasing heat loss'.

6.3.3 Controlling heat loss

Decreasing heat loss

Temporary insulation: Some animals are capable of insulating their body by using temporary methods.

- *Temporary growth of coats:* as explained earlier, permanent insulation is basically a passive thermoregulatory strategy that enables animals to transfer less heat from their bodies to the external environment. The insulation of fur and fat in animals, and feathers in birds is passive as these features are not the result of a change in the body of an animal but instead are present during the organism's lifetime. However, there are exceptions where a species grows one of these insulating mediums to decrease heat loss and heat gain for a temporary period of time. These types of insulation can be recognised as active strategies. For example, some birds grow additional feathers in winter.
- *Waterproofing: Phyllomedusa* is a species of frog which lives in deserts and waterproofs its skin to control heat loss by spreading lipid secretions to cover its limbs. Lipid secretions significantly decrease water loss when basking in sunlight.
- *Preening:* Birds living in Antarctic latitudes use preening as a strategy to reduce heat loss. Through preening, birds transfer oils from their preen gland to their feathers. The oils create a waterproofing layer which traps air next to their bodies, reducing heat loss (Davenport 2012).
- *Piloerection (see 8.3.1.3) and ptiloerection:* These are the "involuntary bristling of hairs or ruffling of feathers; in thermal physiology, an autonomic thermoeffector response often associated with behavioural (e.g., postural) adjustments" (IUPS Thermal Commission 1987). For bird, fluffing their feathers, or for animals, raising their fur, thickens their insulating layer, which helps to reduce heat loss.

Hidromeiosis: This is explained as "the swelling of keratinized layers of the skin, due to prolonged exposure to water or sweat, which blocks sweat ducts and reduces sweating rate" (IUPS Thermal Commission 1987). This is a process that is rapidly reversible (Sargent 1961) and requires that the skin is wet for it to happen, which could happen in a hot, humid climate. Since sweating is a cooling mechanism, blocking it will affect the ability of the body to lose heat in this way.

Body shrinkage: The water bear, or tardigrade, which lives in alpine regions, is able to shrink its body and hence its surface area. The body surface of these animals desiccates to reduce heat loss through their skin during periods of low humidity in their alpine environment (Wharton and Brown 1991).

Countercurrent heat exchange: Many birds and mammals use countercurrent heat exchange to transfer heat between warmer and cooler parts of their bodies. In the legs of some wading birds, for example, there is an artery that conducts warm blood from the heart to the foot. Blood returning from the cold foot ascends through veins, reducing the temperature of blood in the arteries so that blood is at its minimum temperature when it reaches the foot. This process conserves heat to keep the body core warm by reducing heat loss because it reduces the difference in temperature between the feet and the surrounding environment.

Some ectotherms regulate blood flow to the skin in order to conserve heat. For example, when iguanas swim in cold water, they reduce blood flow to their skin, allowing them to retain the heat obtained through basking.

A number of factors affect the amount of heat transferred to and carried through blood, such as blood velocity, heat transfer coefficient, diameter of blood vessels, thickness of blood as well as the temperature of surrounding tissues (Sinha et al. 2016).

Circulatory heat transfer: All endotherms and some ectotherms are able to control heat loss, in addition to generating heat, through physiological means such as dilation and vasoconstriction of blood vessels. This happens during basking in the sun (Davenport 2012). A series of interrelating factors contribute to allowing heat to circulate around an organism's body through its circulatory heat exchange system. These factors are: (1) blood flow in capillaries through superficial body tissue, (2) blood flow through veins and (3) the organism's respiratory ventilation mechanism rate. Essentially, the widening (vasodilation) and contraction (vasoconstriction) increases and reduces heat loss respectively (Davenport 2012).

- ***Vasoconstriction***: Vasoconstriction is the shrinking of the diameter of blood vessels to decrease heat loss. Through this process, blood bypasses an organism's skin surface and follows a different path in returning to the heart. For example, when iguanas dive in water to forage, heat loss is reduced by peripheral vasoconstriction. After diving, iguanas bask in sunlight to compensate for heat loss.

Increasing heat loss

Circulatory heat transfer

- *Vasodilation:* Blood leaves the heart and moves towards capillary beds in the surface of the skin during vasodilation, which increases heat loss. Some mammals are mostly covered in fur, but benefit from a combination of specific blood vessel networks in parts of their bodies that are furless. For example, jackrabbits have large furless ears with extensive networks of blood vessels which facilitate rapid heat loss, enabling them to survive the heat of desert environments.

a) Selective brain cooling: This comes under the vasodilation category and has been defined as "lowering of brain temperature, either locally or as a whole, below arterial blood temperature" (IUPS Thermal Commission 1987). The venous blood will have been cooled by exposure at the surface and on return to the brain, acts as a heat sink.

b) Menopausal hot flush (flash): This has been described as "an abrupt heat dissipation response, occurring in hypoestrogenic women. The response typically consists of peripheral vasodilation, sweating and reports of intense warmth" (IUPS Thermal Commission 1987).

*Evaporative mechanisms: When an animal's body temperature is above ambient temperature, evaporative heat loss is the only means by which heat can be lost (Blatteis 1998). The different types of evaporative heat loss mechanisms are categorised below:

- *Sweating*: Sweating is an autonomic mechanism only seen in mammals in which the skin releases water. Sweating is recognised as the most important heat loss mechanism in humans (Shibasaki and Crandall 2010), where it occurs under sympathetic cholinergic control in which brain processes are affected by cholinergic neutrons, whereas in animals sweating is adrenergically controlled. In humans and animals, heat loss mechanisms have been respectively referred to as somatomotor and vasomotor responses (Blatteis 1998). Thermosensitive neurones send signals to the hypothalamus in response to temperature changes, and the hypothalamus is thought to behave as a set point manager, analogous to a building's thermal management system.

- *Panting*: Many mammals pant which increases evaporative cooling and heat loss. Panting occurs through increasing evaporative loss from the mouth area, which is a low energy thermoregulatory mechanism. Water evaporation from the skin is called cutaneous evaporative heat loss. While both birds and mammals pant, they use different strategies. Some species, such as dogs, combine panting with countercurrent heat exchange, which can prevent their brain from overheating. Panting can be categorised into the following groups:

a) Thermal hyperpnea: "An increase in tidal volume associated with an increase in alveolar ventilation occurring during severe heat stress which has caused a large rise in core temperature" (IUPS Thermal Commission 1987). Thermal hyperpnea in animals, also known as second phase panting, typically involves slow, deep breathing, in contrast to first phase panting, which involves rapid, shallow breathing.

b) Thermal tachypnea (Synonym: thermal polypnea): "A rapid respiratory frequency accompanied by an increase in respiratory minute volume and, commonly, a decrease in tidal volume, in response to a thermoregulatory need to dissipate heat" (IUPS Thermal Commission 1987).

c) Gular fluttering: "...a characteristic form of thermal panting in some birds during exposure to high ambient temperature, by which means air is moved across the moist surfaces of the upper respiratory tract" (IUPS Thermal Commission 1987). The factors that affect the panting mechanisms of birds include the bird's respiratory parameters, properties of the air sac and the cooling efficiency of their panting (Richards 1970). Pelicans, cormorants, turkey vultures, roadrunners, quails and goatsuckers are some of the species the use this behaviour to increase heat loss.

Avoiding seasonal heat loss: Apart from species that live in the tropics, organisms experience different ambient temperatures at different times of the year. Thermoregulatory mechanisms such as hibernation to avoid low temperatures have already been mentioned. Some organisms use the strategies of freeze avoidance and freeze tolerance in extreme cold situations. Both strategies revolve around the need to control the formation of ice. Freeze tolerant organisms preserve their intracellular fluid space from freezing by controlling ice formation in their extracellular fluid spaces. Conversely, surviving intracellular freezing necessitates a freeze avoidance mechanism adaptation strategy, which usually occurs in response to a spontaneous decrease in temperature. In freeze avoidance, organisms reduce the supercooling point of their body fluids to decrease the likelihood of intercellular ice formation, because intercellular crystallisation would be lethal. This mechanism is usually utilised in winter, when organisms experience sudden nucleation.

*Freeze tolerance: While ectotherms are vulnerable to freezing, some species, for example barnacles, mussels, intertidal gastropods and overwintering insects, can survive the freezing of their extracellular body fluids. Many plants also use freezing in their extracellular fluid spaces for survival (Wisniewski et al. 2014).

Freeze avoidance: Freeze avoidance can be either a physiological or biochemical adaptation. Supercooling is recognised as a conforming mechanism. Low temperatures are a key feature of the Antarctic environment. Here, arthropods survive sub-zero temperatures by relying on supercooling as a freeze avoidance strategy. For example, the Antarctic mite *Alaskozetes Arcticus* has a supercooling mechanism which relies on the accumulation of glycerol, together with the purging of ice-nucleating particles from the mite's gut (Young and Block 1980). Further examples include the development of antifreeze proteins in some species of fish (Barnes 2012), and high levels of cryoprotective glycerol in the haemolymph (an arthropod's circulating fluid) of a number of insect species. This same chemical substance is found in the blood of Antarctic Icefish. Although there are a number of examples of physiological and biochemical freeze-avoidance adaptations, most ectotherms employ behavioural adaptation mechanisms in order to avoid freezing (Davenport 2012).

6.4 Controlling heat: thermal adaptation in plants

Initially, a similar approach to that for organisms was taken to classifying the thermal adaptation mechanisms of plants. However, for many plant strategies, the boundary between passive and active strategies was unclear, which restricted their separation into the same categories. Some attempts have been taken to assign them into appropriate categories in Chapter 8.

The range of responses to temperature in plants is extensive from minor morphological and anatomical changes to substantial physiological and biochemical alterations. Morphological thermal adaptation in plants involves various mechanisms such as a reduction in cell size, increased stomata and trichrome densities, closure of stomata and enlarged xylem vessels. Xylems are the vascular tissues that conduct water and nutrients from the roots to the other organs in plants. Physiological changes are substantial and can influence the rate of photosynthesis. Biochemical mechanisms in plants are protein-induced (Waraich et al. 2012), and are beyond the scope of this book.

While both cold and heat stresses affect plants at the scale of the whole organism—multicellular (the organ), cellular and molecular—the impacts of low temperatures on plants are more serious compared to those of high temperatures, as a plant's responses to cold stress involve physiological changes (Shinozaki and Yamaguchi-Shinozaki 2000). Although the biochemical and physiological adaptation processes of plants may not be directly applicable to building design, studying the influences of these processes on the structure and morphology of plants, at both the macro and micro levels, could suggest innovative ideas for architecture.

Plants use one or a combination of morphoanatomical, physiological and biochemical alterations to cope with high temperatures. Furthermore, thermotolerance differs during the lifecycle of a plant. For example, high temperatures could either slow down or stop the seed germination process while the effect of heat stress could be minor for young plants. The intensity of heat stress also affects the thermotolerance of plants (Wahid et al. 2007). Generally, adaptation to heat stress in plants has connections with other processes such as photosynthesis, respiration and regulation of water absorbance. This means that regulating heat can influence gas exchange and water-related processes (Wahid et al. 2007). Examples of these are introduced in Sections 6.4.2 and 6.4.3.

An extremely high temperature leads to the collapse of cellular organisation causing major cellular injuries and even cell death (Wani and Kumar 2020). This can also happen at a moderately high temperature if plant species are exposed to heat stress for a long time. Cell injuries due to high temperatures can thus happen suddenly or slowly with the former called 'direct' heat injury and the latter, 'indirect' heat injury (Waraich et al. 2012). Direct injury leads to denaturation and aggregation of protein along with increased fluidity of membrane lipids. This is a change in physiological processes, while in an indirect heat injury, biochemical alterations occur to modify the developmental patterns (Wani and Kumar 2020).

Plant responses to heat stress range from physiological changes occurring in the bodies of plants to biochemical modifications that cause genetic alterations at molecular levels (Chinnusamy et al. 2007). In addition to accommodating physiological changes at the scale of the whole organism, plants frequently send a signal to change their metabolism through producing a different type of solute that keeps proteins and cellular structures organised but that can slow down the consequences of heat stress (Hasanuzzaman et al. 2013). One of the major consequences of heat stress is an increased number of Reactive Oxygen Species (ROS). ROS are unstable molecules containing oxygen that can easily damage other molecules in a cell through reacting with these.

A combination of physiological and biochemical processes develops heat tolerance in plants. Plants are classified into three groups based on the preferred temperature for their growth: psychrophiles, mesophiles and thermophiles. Plants in the first group grow at low temperatures (such as alpine plants that grow above the tree line), those in the second group favour moderate temperatures (typified by garden plants in temperate climates) and those in the third category can adapt to high temperatures (such as agaves and cacti). Therefore, thermal adaptation in plants varies depending on how much heat stress they can tolerate (Hasanuzzaman et al. 2013).

Figure 6-4 shows the hierarchical classification of thermal adaptation strategies that plants use to control heat.

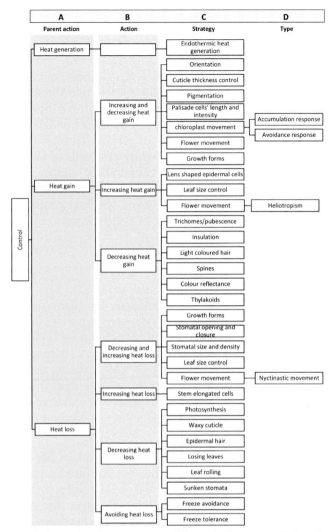

Figure 6-4. The hierarchical categorisation of thermal adaptation solutions for plants.

6.4.1 Generating heat

*Endothermic heat generation: While endothermy is usually associated with endotherms, endothermic heat generation in floral tissues has been known to occur in a wide range of plants (Watling et al. 2008). Endothermy in plants enhances rates of pollination and is associated with floral development, presumably to protect plants from low temperatures. However, endothermic heat generation in this sense has been differentiated from thermoregulation. This is because to be recognised as a thermal

regulating species, plants need to have the capability of sensing the external temperature and regulating heat production at cellular levels (Watling et al. 2008). Only a small number of plants have such capability, such as the sacred lotus.

Another difference can be seen in respiration, as for mammals respiratory processes generate metabolic heat while the extremely low rate of respiration in some plants is not enough for heat generation. Respiration in plants generally takes place in mitochondria and through electron transport to produce energy for ATP (adenosine triphosphate). This is an energy-carrying molecule found in the cells of all living things that powers cellular processes. In endothermic plants, the electron transport is uncoupled from ATP generation and thus, the energy can be used for generating heat. In other words, the heat generation process occurs through an alternative pathway of respiration.

6.4.2 Controlling heat gain

Decreasing and increasing heat gain

Orientation: The environmental challenges for plants in hot deserts are high temperatures and high levels of solar radiation. Tall plants can control light absorbance to some extent by orientating their leaves to an angle which is not perpendicular to the sun's rays (Barnes 2012). Some plants orientate their leaves from east to west, causing them to absorb more sunlight during the early morning and late evening. This accordingly protects the plants from the midday sunlight. The leaves of desert plants are often oriented vertically to reduce leaf dimensions and thus minimise excessive heat load (Curtis et al. 2012, Leigh et al. 2012).

Cuticle thickness control: The cuticle is a transplant film of lipids on the outermost surfaces of tissues in the above ground organs of plants, these being leaves, flowers, seeds and non-woody stems (Riederer 2018). Cuticles in plants act as an interface between the plant and the atmosphere. The primary function of the cuticle is to control permeability through decreasing water evaporation from its outer surface which is covered with wax (Schonherr and Merida 1981). While developing waxy cuticles reduces latent heat loss from the surface of a leaf, there are many contradictory reports of UV-B effects on plant architecture (Robson et al. 2015) and that the thickness of cuticles control solar heat gain. A literature review showed that many studies support the idea that thick cuticles protect the leaves from solar radiation (Haworth and McElwain 2008), which can consequently reduce solar heat gain. Also, it has been argued that leaves tend to be thinner in the absence of solar radiation (Niinemets 2007, 2010), in order to increase solar heat gain for photosynthesis.

Cuticles play a significant role in gas exchange regulation while also protecting the leaves of plants from mechanical injury (Domínguez et al. 2011). Plants control heat gain through modifying the chemical composition and biophysical properties of their cuticles. Recent research shows that some such cuticle modifications have been induced by climate change and the resultant increase in thermal stressors (Heredia-Guerrero et al. 2018).

Relative humidity (RH) and temperature are key factors affecting the biochemical properties of cuticles (Domínguez et al. 2011, Domínguez et al. 2011). The integrity of cuticles as transpiration barriers needs to be maintained, even at extreme temperatures (Schuster et al. 2016). For plants growing in deserts, adaptation to both high temperature and water scarcity is integrated as plants are affected by these environmental stressors at the same time. The relationship between temperature and RH necessitates first understanding how water permeability is controlled by cuticles in extreme temperatures, and then exploring how the permeability level relates to variations in the cuticle thickness.

Evolving a heavier, stiffer and more inert cuticle is one way of adapting to low temperature (Heredia-Guerrero et al. 2018). Inversely, an increase in temperature reduces the stiffness of cuticles. The transpiration rates of cuticles correlate with temperature which means higher temperatures increase the water diffusion coefficient and result in changes in the viscoelasticity of cuticles to enhance water loss (Riederer and Schreiber 2001). While several attempts have been made to define a correlation between water permeability and cuticle thickness (De Micco and Aronne 2012, Smith et al. 2012), it seems that this relationship is not clear (Kerstiens 1996, Riederer and Schreiber 2001).

Even though research supports a negative correlation between both cuticle thickness and the presence of waxes with cuticle water permeability, another study argues that "the frequent statements in textbooks and in the scientific literature that thicker cuticles or cuticles containing more wax are better barriers against transpiration than thin ones with only small amounts of wax are not supported by experimental evidence" (Schuster et al. 2016). The inability to resolve this problem might relate to the complexity of the thermal adaptation processes used by plants.

The structural composition, thickness and quantity of the cuticle varies between plant species, the organs evolving the cuticle and the developmental stage of the plant (Fernández et al. 2016). However, generally, xerophytes or plants that live in dry environments have thick cuticles. For example, the cuticle in succulents has two or three layers of hypodermis. Such xerophytes survive during hot and dry seasons through evolving thick photosynthesis tissues in leaves and stem. These enable plants to limit water loss and store water under the tissues to avoid cell collapse during hot and dry periods. Xerophytes have the most morphological and anatomical diversity and are generally classified into three categories: (1) ephemeral annuals (such as *argemone Mexicana*—the Mexican poppy that completes its lifecycle before the harsh conditions occur), (2) succulents (such as aloe and agave) and (3) non-succulent perennials (such as *calotropis procera*—Apple of Sodom) based on their morphology, physiology and lifecycle pattern. Plants in the first category avoid drought by completing their life cycle within a short period when the humidity level is favourable. In dry conditions, these types of plants create seeds that are highly resistant to aridity and can, therefore, survive.

Flower colouration (pigmentation): The colour variation in flowers can happen due to the regulation of a pigment called anthocyanin (van der Kooi et al. 2019). The temperature of the climate in which a plant grows influences the colour variation. Other environmental factors such as drought stress and exposure to ultraviolet radiation also affect the pigmentation process. This might also mean that pigmented flowers use the strategy of colouration to adapt to heat and drought stress (Arista et al. 2013, Dafni et al. 2020).

Dark flowers may have higher intra-floral temperatures as the solar energy absorption in their petals is more than that absorbed by light-coloured and reflective flowers (Mckee and Richards 1998). Plants growing in low temperature environments evolve darker colours to increase solar absorption. When a flower is closed, the temperature inside the petals also increases due to an underlying physical mechanism in which the internal space operates as a micro-greenhouse capable of absorbing shortwave sunlight and reflecting longwave radiation (Kevan 1989). While there is research supporting these results on colour variation, Shrestha et al. (2018) suggest there is no clear relationship between temperature and colour.

In addition to pigmentation, epidermal cell structures and sub-epidermal air spaces in a leaf are also known to affect the colour variation as they both affect how light is absorbed or reflected, both on the leaf surface and inside it. Given this, the colour variation of leaves can be either pigment-related or structure-related. The latter has variations of two types: 'air space' and 'epidermis'.

> "The 'air-space' is a variation in palisade-cell development, in which the palisade cells exhibit rounded shape with air spaces that produce a pale grayish-green appearance, as can be found in *Begonia* spp. The second type, i.e., 'epidermis' type of variegation, was described in several species, e.g., *Oxalis mariana*, where epidermal attachment to the underlying cell layers varies across a leaf, and sub-epidermal air spaces appear as white or whitish due to their stronger light reflectance" (Shelef et al. 2019).

An assumption is that the light-coloured patches on the surface of a leaf have thermal advantages as the increase in light absorption due to the lens-shaped cells in these areas, create a warmer microenvironment in sub-epidermal spaces (Hüner et al. 2013, Shelef et al. 2019).

Palisade cell length and intensity + chloroplast movement: Light absorption has a positive correlation with heat gain. This means that an increase in the former will lead to an increase in the latter. However, in this strategy, responding to light stress is the primary action that can indirectly affect heat gain in leaves.

Regulating light absorption in plants occurs at the organ, tissue, cellular and organelle levels—all induced by a photoreceptor protein that regulates leaf flattening, stomatal opening, palisade cell development and chloroplast movement (Suetsugu and Wada 2020). A combination of these optimises light utilization for photosynthesis in leaves. This is discussed further in this section.

Palisade tissue is the layer of columnar cells that can be found beneath the upper epidermis in leaves. It has also been called palisade mesophyll, palisade parenchyma and palisade layer (Figure 6-5). Plants grown in an environment with high light intensity contain more palisade cells compared to shade-grown plants (Davis et al. 2011). Columnar palisade cells reduce light scattering and can enable light to penetrate deeper into a leaf; however, the direction and wavelength of the external light affect the efficacy of the columnar palisade layer (Suetsugu and Wada 2020). As shown in Figure 6-5 mesophyll tissue has another layer that contains spongy cells. Such cells can be found under the palisade tissue and can enhance light scattering inside the leaves. This is because there are relatively more cell-air interfaces in the spongy mesophyll layer compared to those in the palisade tissue (Vogelmann and Martin 1993).

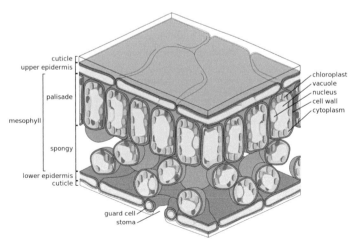

Figure 6-5. Leaf structure (Zephyris 2011).

Palisade tissue cells are rich in chloroplasts that exhibit movement in response to light. Chloroplast movement affects the optical properties of leaves and hence, adjustment to light transmission. Plants growing in shadow exhibit no or very weak chloroplast movement. This might be because shade-grown plants rarely receive the excess light that would need to be transmitted rather than being absorbed.

Number, size and arrangement are the three anatomical features of chloroplast that can affect the internal light environment in a leaf at the intracellular level (Xiao et al. 2016). However, it seems that changes in the optical properties of leaves induced by chloroplast relocation are highly variable between species (Davis et al. 2011). There are two main patterns for chloroplast movement that adjust light absorption and transmission, and thus enable increasing and decreasing heat gain in leaves (Figure 6-6). In low-light conditions, plants use the 'accumulation response' in which chloroplasts accumulate along the upper and lower surfaces of cells to increase light absorption for photosynthesis and reduce transmission (Davis et al. 2011). Accordingly, shade-grown leaves have shorter cells with larger diameters to increase the potential for chloroplast movement. Also, the palisade layer in shade-grown leaves is usually single as most of the sunlight is used in the first palisade layer and would not even reach beyond it. Conversely, plants have evolved another mechanism to avoid excess light absorption in the face of high light intensity, called the 'avoidance response'. This occurs through the movement of chloroplast to the cell walls to avoid the incident light. The change occurs at the cellular level so the tissue level remains unchanged (Kasahara et al. 2002).

The relationship between palisade shapes, chloroplast movement and photosynthesis has been investigated. It has been suggested that the

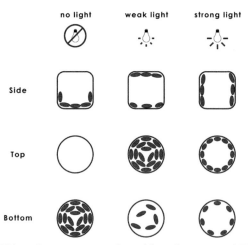

Figure 6-6. Chloroplast movement, adapted from Suetsugu and Wada (2020).

increase in the diameter of cells enables the 'accumulation response' in shade-grown leaves and thus a compact palisade tissue packed with a large number of columnar cells with smaller diameters could limit the chloroplast movement (Davis et al. 2011). More recent research showed a contrasting result by explaining that it is the cell shape that determines how much chloroplast can move (Gotoh et al. 2018), which is why plants living in climates with high light intensity develop longer columnar cells. The longer cell has a larger wall area to accommodate chloroplasts. Furthermore, an adjustment in the thickness of leaves modifies the rate of photosynthesis and is thus important for plants (Terashima et al. 2005). However, the mechanism behind leaf thickening is still unclear as recent research has proposed various influential factors such as either only cell elongation in the palisade layer, or the number of cell layers in the mesophyll tissue, or a combination of both (Hoshino et al. 2019).

* Flower movement

- *Nyctinastic movement (opening and closing behaviour)*: The opening and closing behaviour of a flower is called nyctinastic movement (see also 6.4.3), and this can protect the reproductive organs from excessive heat or cold (van Doorn and van Meeteren 2003, van der Kooi et al. 2019).

Growth forms: Desert plants evolve compact forms with tightly packed leaves to reduce solar absorption through reducing exposed surface area (Solbrig and Orians 1977). Among the growth forms of cacti, barrel forms are the most heat-tolerant, followed by the columnar and finally, the globose forms (Reyes-Olivas et al. 2002). Detailed studies on the morphological characteristics of cacti have been conducted (Reyes-Olivas et al. 2002, López and Valdivia 2007, Bobich and North 2009) but are out of the scope of this book. In polar and mountain environments, short plants benefit from absorbing the heat radiated from the soil surface. Decumbent, rosette and cushion are the growth forms frequently found in such environments (Fernández-Marín et al. 2020).

Increasing heat gain

Lens-shaped epidermal cells: A type of epidermal cell found in plants has a lens-shaped structure that is assumed to have been evolved to focus light within the upper layers of a leaf. It has been hypothesised that such cells focus the incident light when it is not perpendicular to the leaf surface and as a result, create a high photon flux intensity in these regions. However, there seem to be contrasting ideas about the development of these cells (Shelef et al. 2019). One study has theorised that where these cells exist, the colour of the leaf is lighter. In that sense, developing lens-shaped cells is as an adaptive response to the poor light conditions of shade-grown plants so that the rate of photosynthesis can increase through the efficient capture of solar energy (Bone et al. 1985). This would mean more light absorption would increase heat gain in leaves. However, another study postulated that lens-shaped cells do not affect diffused light capture (Brodersen and Vogelmann 2007).

Leaf size control: There are contrasting ideas about whether shade-grown plants have larger leaves to enhance light capture. While some studies support this idea (Niinemets 2007, Martin et al. 2020), other studies have proposed that large leaves are uncommon features in cold environments as leaves can be damaged by frost during the night. It seems that the relationship between leaf size and climate, as suggested by Wright et al. (2017) (who analysed leaf data for 7670 plant species), needs to go beyond the existing simple patterns.

* *Flower movement*

- *Heliotropism*: Heliotropism or solar tracking is another type of movement when plants turn towards a directional heat source. The difference between orientation (see 6.4.2) and heliotropism is that in the latter, the organ's orientation is adjusted continuously over a day. While several studies support the idea that heliotropism increases floral temperature, there has been a recent mention of the need for more research to confirm the theory (van der Kooi et al. 2019). This has been recognised as important due to the complexity of thermoregulation in plants and the fact that other factors such as size, form, colour, phylogenetic ancestry and geographical distribution of species affect the plant's temperature.

Decreasing heat gain

* *Trichomes/pubescence*: A trichome is a fine outgrowth like a hair or scale found on some plants. Trichomes are found on different organs in plants such as leaves, buds and roots. They have fine structures and are made of one or several cells that protrude from the epidermis. While the primary function of trichomes is absorbing moisture, there has been discussion of the role of trichomes in controlling solar heat gain. It has been assumed that the morphology of trichomes regulates leaf absorption of radiation for alpine plants (Mershon et al. 2015). It has also been suggested that an increase in trichomatous density leads to reducing solar absorption as trichomes can function as a shield covering the plant tissue (Benzing and Bennett 2000, Dutta et al. 2020). This subsequently decreases heat gain (Konrad et al. 2015).

Root trichomes help in absorbing nutrients and water. In some species such as bromeliads, trichomes on the surface of the leaves fully replace the absorbing function of the roots (Benzing et al. 1985, Leroy et al. 2019).

Also, some vascular epiphytes that use other plants as their hosts to grow do not have access to the soil and thus use the trichomes on their leaves for rainwater and nutrient uptake (Gotsch et al. 2015). Some studies have argued that the highly reflective substance secreted from glandular trichomes reduces water loss (Wagner 1991, Lenssen et al. 2001, Rodriguez et al. 2018), while another study linked water loss with the high density of trichomes (Abd Elhalim et al. 2016). Reducing water loss leads to reduced latent heat loss.

Insulation: Tree bark, especially the outer bark which is composed of dead cells, serves an insulating purpose in fire-prone areas to protect the inner tree and thus allow it to resume functioning once the fire is over (Lawes et al. 2011, Bär and Mayr 2020). This has a direct parallel in building design where timber structural members can be oversized as a fire protection action. In a fire, the outer surfaces char and the layer of charred wood insulates the inner core from further burning. With sufficient residual unburned timber, the collapse of the timber structure can be delayed (Aseeva et al. 2014).

Light coloured hair: While obvious, light-coloured trichomes reflect light and thus reduce solar absorption.

Spines: Spines function as shading surfaces that reduce solar heat gain in desert plants (Drezner 2011, Wickens 2013). Also it has been suggested that spines are able to collect the moisture in the air during the night to increase evapotranspiration during the day (Hegazy and Doust 2016).

Colour reflectance: The level of reflectance of leaves also affects their light absorbance. Hairs and wax on the surface of leaves can provide a barrier to light absorption. Whitish, silver and greyish colours are other evolutionary adaptation mechanisms developed by plants in hot deserts (Holmes and Keiller 2002).

Thylakoids: The chloroplast contains membrane sacs called thylakoids which are arranged into stacks known as grana (Figure 6-7). The structural changes taking place in the thylakoid membrane system are light-induced. Thylakoids contain chlorophylls and are the place where light-dependent reactions occur. Plants alter their structure at the cellular level to avoid excess sunlight that can affect photosynthesis, as high light intensities produce a toxic product in stomata and relevant photosynthesis apertures (Kirchhoff 2014). Responding to light stress happens through a cycle of stacking and unstacking the thylakoids which is called the damage and repair cycle.

Light-harvesting proteins called PSII are embedded in the lipid layers of thylakoids and can produce energy through electron transport. The total surface area for electron transport is larger in stacked than in

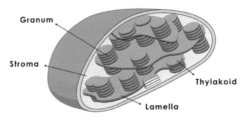

Figure 6-7. Chloroplast, adapted from (Miguelsierra 2011).

unstacked thylakoids, such that "...thylakoid unstacking is a mechanism whereby plants protect PSII from photo-oxidative damage under high light" (Yoshioka-Nishimura 2016). However, a recent study has proposed an alternative to the commonly accepted short-term regulatory process of thylakoids (Johnson and Wientjes 2020).

6.4.3 Controlling heat loss

Decreasing and increasing heat loss

*Growth forms: The compact growth form of cushion plants in dry regions reduces airflow over the surface of the epidermis. This reduces the rate of water loss through decreasing evaporative heat loss. Conversely, succulent stems swell during the rainy season to store water for the drought season so they can use the collected water for transpiration cooling (Mauseth 2000). Taller swollen ribs expand more that the short ones and thus will have a larger surface to volume ratio, which significantly enhances transpiration.

Generally, it seems that the size and height of plants are restricted where temperatures are extreme and where water is restricted. Low growth reduces exposure to wind, which can dry out and damage the plants. It also shortens the distance between the soil and the leaves, which helps the water balance, and helps deal with low temperatures by limiting exposure. "Early ecophysiological field studies have shown that short-stature plants benefit in cold environments from the higher temperatures produced by radiative heating of the soil surface, strongly decoupling their foliage temperature from the air temperatures" (Fernández-Marín et al. 2020). Given this, the first two reasons for low growth are concerned with reducing heat loss while the latter increases heat gain through radiation.

*Stomatal opening and closure: Stomata are pores in the epidermis of the leaf or stem of a plant. The pores are like a slit that can open and close, and so control the movement of gases into the plant (like oxygen for respiration and carbon dioxide for photosynthesis) and out (like water vapour). At low temperatures, plants close their stomata to decrease water loss and subsequently heat loss through the cooling effect of water evaporation. When the environmental temperature increases, stomata open to induce

water evaporation. However, plants balance water as water loss might lead to drought damage, which is known as hydraulic failure. While plants typically respond to heat stresses by balancing internal temperature and water capacity, the order in which they do these seems to vary in different conditions. This has been referred to as an ecological trade-off. For example, a research study was conducted to observe how the leaves of 14 species of grasses control heat flux in the face of drought (Maricle et al. 2007). Some species were native to high marshes while others were specific to low-lands. The results show all species were mostly dependent on latent heat flux which is the heat loss from water evaporation through the leaves. However, when the available water reduced, the latent heat flux dropped by 65% to stop loss of water while this increased the temperature of the leaves by 4°C. Sensible heat loss through conduction and convection was higher for high marsh species.

Considering that stomata adjustment takes from 2 to 60 minutes, in the event of a sudden increase in solar radiation, plants use the water content of their leaves for evaporative cooling in order to avoid overheating (Schymanski et al. 2013). This happens as a rapid and extreme fluctuation in irradiation can occur in a few seconds, which is not enough time for a plant to use stomata opening to increase respiration.

Avoiding excess water evaporation is complicated and seems to be dependent on how leaves are connected to xylems. Evaporation of water in the leaves stops when the water pressure reduces significantly. The drop in the pressure occurs in tissues with low conductivity. Xylems can be either (1) separated from leaf tissues by a low conductivity barrier, (2) linked to the epidermis or (3) linked to all the leaves. Accordingly, to avoid excess evaporation through stomata closure, three scenarios happen in plants:

> "for case 1, the pressure drop would occur between the leaf xylem and all other tissues, i.e. stomata would close autonomously when the entire leaf tissue reaches a critical water depletion and potential, whereas the leaf xylem potential would remain relatively unchanged. In case 2, autonomous stomatal closure would be expected when water depletion in the epidermis becomes critical, while the mesophyll can maintain higher water potential. In this case, the leaf xylem potential would be expected to decline together with the water potential of the epidermis. In case 3, like in case 1, all leaf tissues would reach a critical water depletion and potential before autonomous stomatal closure, but in this case, the leaf xylem potential would also decline" (Schymanski et al. 2013).

Figure 6-8 shows a diagram of the water column from soil to leaf cells.

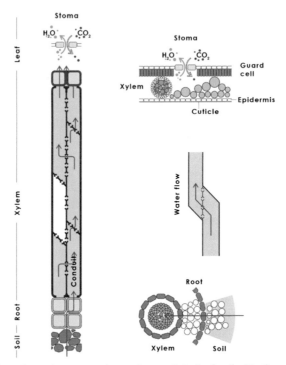

Figure 6-8. The schematic water column from soil to leaf cells (Author, adapted from Venturas et al. (2017)).

Stomatal density (SD) and stomatal size (SS): The function of a particular stomata is determined by its morphological characteristics among which density and size have direct influences. Furthermore, the morphological features of stomata differ between species (Figure 6-9). When the environmental temperature increases, some plant species can produce leaves with an altered stomatal density which affects evaporative cooling and subsequently, transpiration rates (Crawford et al. 2012, Jumrani et al. 2017). However, it is not only the temperature that determines stomatal density, but also other environmental factors such as light intensity, water availability and oxygen level (Sekiya and Yano 2008, Khazaei et al. 2013). Given this, it is argued that "variation in size and density of stomata may arise due to genetic factors and/or growth under different environmental conditions" (Bertolino et al. 2019).

The short-term alteration of SD and SS will affect stomatal conductance of both water and CO_2, which will lead to decreasing or increasing heat loss. For example, research shows that stomatal density increases with a rise in temperature (Zhu et al. 2018), whereas the reverse happens with a decrease in temperature. While a short term rise in temperature increases stomatal density, stomatal size is more likely to decrease, as the remaining

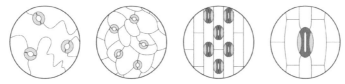

Figure 6-9. Stomatal traits in different species (Author, adapted from Bertolino et al. (2019)).

space on the surface of the leaf is necessary for other processes such as photosynthesis (Hill et al. 2015).

Not all research has produced the same results when it comes to explaining the effects of global warming as a long-term stress on stomatal size and density. While SD and SS are significantly affected by warming, it is not clear whether these parameters increase, decrease or remain constant. As Wu et al. (2018) state:

> "…in contrast to increases of leaf stomatal density in response to warming, decreases in stomatal density were also observed. Besides, some studies showed no change in stomatal density upon warming treatment. Similarly, warming had diverse effects on stomatal size across species, such that either increases, decreases, or no change in stomatal size were reported".

Leaf size control: The size and shape of a leaf play a significant role in thermal regulation compared to other morphological features of the plant. These proportions govern the leaf temperature through an air layer around it. In this boundary layer, heat transfer is less compared to the air movement beyond it. Figure 6-10 shows a schematic cross-section of a leaf.

> "All other things being equal, the thickness of a leaf boundary layer increases with distance from the windward edge and therefore with leaf size, such that the rate of heat convection per unit area is greater between leaf and air for small leaves than large leaves. This leads to equilibrium temperatures closer to the air for small than large leaves and is the most widely accepted explanation for the presence of smaller leaves in regions such as deserts" (Leigh et al. 2017).

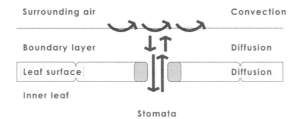

Figure 6-10. Cross-section of a leaf (Author).

Given this, the small size of leaves is an effective thermal adaption strategy for plants growing in hot environments while the leaf shape probably has multiple evolutionary drivers of which temperature control is not a prime one (Leigh et al. 2017). Variation in leaf shape appears to be constrained by genetics as certain shapes occur independently of climate (Jordan 1997). Some researchers argue that for species in hot and dry environments, leaf dissection, or the way in which a leaf is divided into lobes and leaflets, does not represent a primary adaptation to the thermal environment. "Where species in hotter environments do possess dissected leaves, improved thermal regulation may be simply a fortunate by-product of other evolutionary drivers of dissection such as improved hydraulic efficiency or reduced solar interception" (Leigh et al. 2017). Another study rejects the idea that temperature is the primary selective driver in the evolution of leaf shape complexity (Li et al. 2016). However, the small size of leaves has other benefits for desert plants. For example, to reduce or stop water loss, succulents appear to have reduced the size of their leaves or modified them into thorns to reduce transpiration. Examples are *Opuntia* sp. (prickly pear), *Euphorbia Splendens* (crown of thorns) and various types of cacti, and agave. Succulents are also known as drought-enduring xerophytes as they use stem, leaves and roots to store water during hot and dry periods. Non-succulent perennials as another category of xerophytes also benefit from having smaller leaves.

* Flower movement

- **Nyctinastic movement (opening and closing behaviour):** In flowers, interior flower temperature is thought to be influenced by other factors such as the movement of flower petals through the opening and closing behaviour, especially for species living in cold regions. Even though this behaviour is mainly used in response to light intensity and humidity, it can affect the temperature inside flowers when closed (Mu et al. 2010). This strategy can be seen in families such as *Gentianaceae* (the Gentian family), and *Iridaceae* (a family of plants including irises and crocuses) (Andrews 1929).

Increasing heat loss

Stem elongated cells: A plant's stem consists of nodes from which the leaves and branches grow. The distance between each pair of nodes is an internode, a growth which requires cell division and elongations in the direction of the growth. Some species of plants such as *Arabidopsis* (rock cress) respond to heat stress by increasing growth through the elongation of stems. One of the influential factors in determining the effect of heat stress on a plant's architecture seems to be the difference between day and night temperatures (Patel and Franklin 2009). Greater elongation happens when the temperature in the day is higher than that at night.

At high temperatures, shade-intolerant plants living close to other plants compete for light, so they increase growth to reach the sunlight. This strategy is called the shade avoidance syndrome. Given this, "it has been speculated that the phenotypically similar elongation responses to high temperature may aid heat dissipation through raising leaf and meristematic tissues towards cooling breeze" (Patel and Franklin 2009). However, stem elongation stops with water scarcity.

Decreasing heat loss

Photosynthesis: Photosynthesis, the process by which plants convert carbon dioxide and water into nutrients using light, is one of the most sensitive physiological processes and reducing this is a means of responding to heat stress. Higher or vascular plants living in thermally contrasting habitats exhibit different photosynthetic responses to heat stress (Berry and Bjorkman 1980). Accordingly, this has an impact on their ability to function at extreme temperatures.

A comprehensive analysis of photosynthesis under heat stress showed that plants respond to heat stress by decreasing the rate of photosynthesis (Berry and Bjorkman 1980), which seems to be caused by changes occurring in the stomatal conductance and intercellular CO_2 concentration. "Stomatal conductance...is the rate of CO_2 entering, or water vapour exiting through stomata" (Thomas et al. 2016). In fact, the photosynthetic response is complicated as it contains several interconnected biological processes.

Photosynthesis uses carbon dioxide and water and is thus dependent on them which means any changes in these contributing factors have an effect. This has been well explained by Song et al. (2014) who note that if stomatal conductance and CO_2 concentration decrease simultaneously, the former will be the influential factor in reducing the rate of photosynthesis, whereas if stomatal conductance increases but CO_2 concentration decreases, the drop in photosynthesis is caused by non-stomatal factors. Finally, if the heat stress exposure continues, stomatal conductance, CO_2 concentration and the rate of photosynthesis decrease at the same time.

In all the cases mentioned above, decreasing photosynthesis is associated with water evaporation from leaves leading to a resultant latent heat loss.

In low temperatures, plants use the strategy of reducing the photosynthesis rate to survive.

Waxy cuticle: (See 6.4.2)

Epidermal hair: (See trichomes, Section 6.4.2)

Losing leaves: Leaf loss is a defence mechanism used by plants when experiencing thermal stressors. A high temperature affects the hydraulic conductance in a leaf and this leads to dehydration and finally, leaf drop.

When the temperature increases, plants enhance leaf evaporation by increasing the transpiration rate. In places with low water availability, evaporative cooling will lead to complete water loss in a leaf if it continues for a long period. An example of a plant using this defence mechanism is *Platanus x acerifolia* (the London plane tree) that can lose up to half of its leaves in a few days to reduce water evaporation from the stomata, which subsequently decreases the transpiration cooling effects in the leaves (Sanusi and Livesley 2020).

Leaf rolling: Leaf rolling is known as a response to environmental stressors including temperature. It seems reducing latent heat loss through evaporation is not the main reason for this mechanism as is it argued it protects leaves from the damage caused by high light intensity (Kadioglu et al. 2012). Leaf rolling reduces transpiration by reducing the exposed area of the leaves. As a result, the leaf rolling mechanism can potentially be used when plants experience a water deficit. Plants that are desiccation tolerant use the leaf curling strategy. To survive desiccation in high light intensity, plants need to possess flexible cell walls. When desiccation happens, the contents of leaves are drawn inwards. As a result, the cell walls come under an extensive pressure to fold inwards, although this can be reversed during rehydration. This is a reversible mechanism that requires flexibility in the cell walls to be effective (Chen et al. 2020).

This mechanism is common in many xerophytes such as *Poaceae* (grasses), *Juncaceae* (the rush family), and *Cyperaceae* (sedges) but can also be found in species experiencing drought and salinity stress (Grigore and Toma 2010). In many xerophytic grasses, leaves roll to reduce the leaf surface in order to restrict loss of water due to transpiration. Examples of other plants that do this are *Calotropis procera* (Apple of Sodom) and *Acacia Nelotica* (thorny acacia).

Resurrection plants, such as the Rose of Jericho, are desiccation-tolerant and have been studied for their ability to unroll during the rehydration period. Experiments have shown that they have the ability to recover. For example, when this happens to *Ramonda myconi* (the Pyrenean violet), "the upward bent peripheral leaves reverted to their initial positions (parallel to the soil surface), thereby exposing the photosynthetically active tissues and the central vegetation point" (Kampowski et al. 2018).

Plants that have the ability to keep their water content in a stable state regardless of the external environment are known as homoiohydric and they can survive in short and moderate dry seasons. Most vascular plants are homoiohydric. While homoiohydric species keep their chlorophylls during a drought, other species like resurrection plants are known as poikilohydric. They degrade their chloroplasts by changing them into desiccoplasts to survive long periods of severe dehydration (Kampowski et al. 2018).

Sunken stomata: Stomata can be located on both the upper and lower epidermis of a leaf. Being sunken into the epidermis of a leaf creates a microclimate for stomata that leads to a decreased evaporation rate.

Avoiding heat loss: Like microbes, plants use one or both of the two cold temperature survival strategies of freeze avoidance and freeze tolerance. The mechanism of the former involves a chemical substance similar to antifreeze, whereas freeze tolerance includes a controlled ice-creation mechanism to prevent intracellular freezing.

Freeze avoidance: (see avoiding heat loss in active methods of thermal adaptation for animals)

Freeze tolerance: (see avoiding heat loss in active methods of thermal adaptation for animals)

6.5 Generating the ThBA

This chapter has described the research on thermoregulation in both plants and animals that was used to generate the first step in a system for finding strategies in nature, known as the ThBA. The next chapter describes how these strategies are linked to buildings, which forms the second step in generating the ThBA.

References

Abd Elhalim, M. E., Abo-Alatta, O. K., Habib, S. and Abd Elbar, O. H. (2016). The anatomical features of the desert halophytes Zygophyllum album LF and Nitraria retusa (Forssk.) Asch. *Annals of Agricultural Sciences, 61*(1), 97–104.

Andrews, F. (1929). The effect of temperature on flowers. *Plant Physiology, 4*(2), 281.

Arends, A., Bonaccorso, F. J. and Genoud, M. (1995). Basal rates of metabolism of nectarivorous bats (Phyllostomidae) from a semiarid thorn forest in Venezuela. *Journal of Mammalogy, 76*(3), 947–956.

Arista, M., Talavera, M., Berjano, R. and Ortiz, P. L. (2013). Abiotic factors may explain the geographical distribution of flower colour morphs and the maintenance of colour polymorphism in the scarlet pimpernel. *Journal of Ecology, 101*(6), 1613–1622.

Aseeva, R. M., Serkov, B. and Sivenkov, A. (2014). *Fire Behavior and Fire Protection in Timber Buildings*. Dordrecht, Heidelberg, New York: London: Springer.

Bär, A. and Mayr, S. (2020). Bark insulation: Ten Central Alpine tree species compared. *Forest Ecology and Management, 474*, 118361.

Barnes, D. K. A. (2012). Polar marine ecosystems. pp. 36–61. *In*: E. Bell (ed.). *Life at Extremes: Environments, Organisms, and Strategies for Survival*. Tarxien, Malta: CABI.

Benzing, D. H., Givnish, T. J. and Bermudes, D. (1985). Absorptive trichomes in Brocchinia reducta (Bromeliaceae) and their evolutionary and systematic significance. *Systematic Botany*, 81–91.

Benzing, D. H. and Bennett, B. (2000). *Bromeliaceae: Profile of an Adaptive Radiation*. Cambridge University Press.

Berry, J. and Bjorkman, O. (1980). Photosynthetic response and adaptation to temperature in higher plants. *Annual Review of Plant Physiology, 31*(1), 491–543.

Bertolino, L. T., Caine, R. S. and Gray, J. E. (2019). Impact of stomatal density and morphology on water-use efficiency in a changing world. *Frontiers in Plant Science, 10*, 225.

Blatteis, C. M. (1998). *Physiology and Pathophysiology of Temperature Regulation.* Jurong East, Singapore: World Scientific.

Block, B. A., Dewar, H., Blackwell, S. B., Williams, T. D., Prince, E. D. and Farwell, C. J. (2001). Migratory movements, depth preferences, and thermal biology of Atlantic bluefin tuna. *Science, 293*(5533), 1310–1314. doi:10.1126/science.1061197.

Bobich, E. G. and North, G. B. (2009). Structural implications of succulence: architecture, anatomy, and mechanics of photosynthetic stem succulents, pachycauls, and leaf succulents. *Perspectives in Biophysical Plant Ecophysiology, A Tribute to Park S. Nobel*, 3–38.

Bone, R. A., Lee, D. W. and Norman, J. (1985). Epidermal cells functioning as lenses in leaves of tropical rain-forest shade plants. *Applied Optics, 24*(10), 1408–1412.

Brischoux, F., Bonnet, X. and Shine, R. (2009). Kleptothermy: an additional category of thermoregulation, and a possible example in sea kraits (Laticauda laticaudata, Serpentes). *Biology Letters, 5*(6), 729–731. doi:10.1098/rsbl.2009.0550.

Brodersen, C. R. and Vogelmann, T. C. (2007). Do epidermal lens cells facilitate the absorptance of diffuse light? *American Journal of Botany, 94*(7), 1061–1066.

Brown, C. R. and Foster, G. G. (1992). The thermal and energetic significance of clustering in the Speckled Mousebird, Colius-Striatus. *Journal of Comparative Physiology B-Biochemical Systemic and Environmental Physiology, 162*(7), 658–664. doi:Doi 10.1007/Bf00296648.

Burda, H., Šumbera, R. and Begall, S. (2007). Microclimate in burrows of subterranean rodents—revisited. pp. 21–33. *In*: S. Begall, H. Burda and C. E. Schleich (eds.). *Subterranean Rodents: News from Underground.* Germany, Berlin: Springer.

Campana, S. E., Dorey, A., Fowler, M., Joyce, W., Wang, Z., Wright, D. and Yashayaev, I. (2011). Migration pathways, behavioural thermoregulation and overwintering grounds of blue sharks in the Northwest Atlantic. *PLoS One, 6*(2), e16854. doi:10.1371/journal.pone.0016854.

Canals, M., Rosenmann, M. and Bozinovic, F. (1989). Energetics and geometry of huddling in small mammals. *Journal of Theoretical Biology, 141*(2), 181–189.

Chaise, L. L., McCafferty, D. J., Krellenstein, A., Gallon, S. L., Paterson, W. D., Théry, M., Ancel, A. and Gilbert, C. (2019). Environmental and physiological determinants of huddling behavior of molting female southern elephant seals (Mirounga leonina). *Physiology & Behavior, 199*, 182–190.

Chen, P., Jung, N. U., Giarola, V. and Bartels, D. (2020). The dynamic responses of cell walls in resurrection plants during dehydration and rehydration. *Frontiers in Plant Science, 10*, 1698.

Chinnusamy, V., Zhu, J., Zhou, T. and Zhu, J.-K. (2007). Small RNAs: big role in abiotic stress tolerance of plants. pp. 223–260. *In*: *Advances in Molecular Breeding Toward Drought and Salt Tolerant Crops.* Springer.

Crawford, A. J., McLachlan, D. H., Hetherington, A. M. and Franklin, K. A. (2012). High temperature exposure increases plant cooling capacity. *Current Biology, 22*(10), R396–R397.

Crawford, R. M. (2013). *Tundra-taiga Biology.* Oxford, UK: Oxford University Press.

Cui, Y., Gong, H., Wang, Y., Li, D. and Bai, H. (2018). A thermally insulating textile inspired by polar bear hair. *Advanced Materials, 30*(14), 1706807.

Currie, A. (2017). Hot-blooded gluttons: Dependency, coherence, and method in the historical sciences. *The British Journal for the Philosophy of Science, 68*(4), 929–952.

Curtis, E. M., Leigh, A. and Rayburg, S. (2012). Relationships among leaf traits of Australian arid zone plants: alternative modes of thermal protection. *Australian Journal of Botany, 60*(6), 471–483.

Dafni, A., Tzohari, H., Ben-Shlomo, R., Vereecken, N. J. and Ne'eman, G. (2020). Flower colour polymorphism, pollination modes, breeding system and gene flow in Anemone coronaria. *Plants, 9*(3), 397.

Davenport, J. (2012). *Environmental Stress and Behavioural Adaptation.* Sydney, Australia: Springer.

Davis, P. A., Caylor, S., Whippo, C. W. and Hangarter, R. P. (2011). Changes in leaf optical properties associated with light-dependent chloroplast movements. *Plant, Cell & Environment, 34*(12), 2047–2059.

Dawson, T. J. and Maloney, S. K. (2017). Thermal implications of interactions between insulation, solar reflectance, and fur structure in the summer coats of diverse species of kangaroo. *Journal of Comparative Physiology B, 187*(3), 517–528.

De Micco, V. and Aronne, G. (2012). Morpho-anatomical traits for plant adaptation to drought. pp. 37–61. *In: Plant Responses to Drought Stress.* Springer.

Dean, W. R. J. and Milton, S. (1999). *The Karoo: Ecological Patterns and Processes.* Cambridge, UK: Cambridge University Press.

DeWitt, C. B., McGinnis, S. M. and Dickson, L. L. (1967). Behavioral thermoregulation in the desert iguana. *Science, 158*(3802), 809–810.

Diego-Rasilla, F. J., Pérez-Mellado, V. and Pérez-Cembranos, A. (2017). Spontaneous magnetic alignment behaviour in free-living lizards. *The Science of Nature, 104*(3-4), 13.

Dittbrenner, N., Lazzara, R., Köhler, H.-R., Mazzia, C., Capowiez, Y. and Triebskorn, R. (2008). Heat tolerance in Mediterranean land snails: histopathology after exposure to different temperature regimes. *Journal of Molluscan Studies, 75*(1), 9–18.

Domínguez, E., Cuartero, J. and Heredia, A. (2011). An overview on plant cuticle biomechanics. *Plant Science, 181*(2), 77–84.

Domínguez, E., Heredia-Guerrero, J. A. and Heredia, A. (2011). The biophysical design of plant cuticles: an overview. *New Phytologist, 189*(4), 938–949.

Doyle, S., Cabot, D., Walsh, A., Inger, R., Bearhop, S. and McMahon, B. J. (2020). Temperature and precipitation at migratory grounds influence demographic trends of an Arctic-breeding bird. *Global Change Biology, 26*(10), 5447–5458.

Drezner, T. D. (2011). Cactus surface temperatures are impacted by seasonality, spines and height on plant. *Environmental and Experimental Botany, 74*, 17–21.

Dutta, P., Chakraborti, S., Chaudhuri, K. M. and Mondal, S. (2020). Physiological responses and resilience of plants to climate change. pp. 3–20. *In: New Frontiers in Stress Management for Durable Agriculture.* Springer.

Eto, T., Sakamoto, S. H., Okubo, Y., Koshimoto, C., Kashimura, A. and Morita, T. (2014). Huddling facilitates expression of daily torpor in the large Japanese field mouse Apodemus speciosus. *Physiology & Behavior, 133*, 22–29.

Fernández-Marín, B., Gulías, J., Figueroa, C. M., Iñiguez, C., Clemente-Moreno, M. J., Nunes-Nesi, A., Fernie, A. R., Cavieres, L. A., Bravo, L. A., García-Plazaola, J. I. and Gago, J. (2020). How do vascular plants perform photosynthesis in extreme environments? An integrative ecophysiological and biochemical story. *The Plant Journal, 101*(4), 979–1000.

Fernández, V., Guzmán-Delgado, P., Graça, J., Santos, S. and Gil, L. (2016). Cuticle structure in relation to chemical composition: re-assessing the prevailing model. *Frontiers in Plant Science, 7*, 427.

Frey, R. W., Howard, J. D. and Pryor, W. A. (1978). Ophiomorpha—its morphologic, taxonomic, and environmental significance. *Palaeogeography Palaeoclimatology Palaeoecology, 23*(3-4), 199–229. doi:Doi 10.1016/0031-0182(78)90094-9.

Ganot, Y., Dragila, M. I. and Weisbrod, N. (2012). Impact of thermal convection on air circulation in a mammalian burrow under arid conditions. *Journal of Arid Environments, 84*, 51–62. doi:10.1016/j.jaridenv.2012.04.003.

Garrick, D. (2008). Body surface temperature and length in relation to the thermal biology of lizards. *Bioscience Horizons, 1*(2), 136–142.

Ghassemi Nejad, J., Kim, B. W., Lee, B. H. and Sung, K. I. (2017). Coat and hair color: hair cortisol and serotonin levels in lactating Holstein cows under heat stress conditions. *Animal Science Journal, 88*(1), 190–194. doi:10.1111/asj.12662.

Gilbert, C., McCafferty, D., Le Maho, Y., Martrette, J. M., Giroud, S., Blanc, S. and Ancel, A. (2010). One for all and all for one: the energetic benefits of huddling in endotherms. *Biological Reviews, 85*(3), 545–569.

Gotoh, E., Suetsugu, N., Higa, T., Matsushita, T., Tsukaya, H. and Wada, M. (2018). Palisade cell shape affects the light-induced chloroplast movements and leaf photosynthesis. *Scientific Reports, 8*(1), 1–9.

Gotsch, S. G., Nadkarni, N., Darby, A., Glunk, A., Dix, M., Davidson, K. and Dawson, T. E. (2015). Life in the treetops: ecophysiological strategies of canopy epiphytes in a tropical montane cloud forest. *Ecological Monographs, 85*(3), 393–412.

Griffing, A. H., Gamble, T. and Bauer, A. M. (2020). Distinct patterns of pigment development underlie convergent hyperpigmentation between nocturnal and diurnal geckos (Squamata: Gekkota). *BMC Evolutionary Biology, 20*, 1–10.

Grigore, M.-N. and Toma, C. (2010). A proposal for a new halophytes classification, based on integrative anatomy observations. *Muz. Olteniei, Craiova, Stud. şi Com., Şt. Nat., 2010c, 26*(1), 45–50.

Grizante, M. B., Brandt, R. and Kohlsdorf, T. (2012). Evolution of body elongation in gymnophthalmid lizards: relationships with climate. *PLoS One, 7*(11), e49772.

Hasanuzzaman, M., Nahar, K., Alam, M., Roychowdhury, R. and Fujita, M. (2013). Physiological, biochemical, and molecular mechanisms of heat stress tolerance in plants. *International Journal of Molecular Sciences, 14*(5), 9643–9684.

Haworth, M. and McElwain, J. (2008). Hot, dry, wet, cold or toxic? Revisiting the ecological significance of leaf and cuticular micromorphology. *Palaeogeography, Palaeoclimatology, Palaeoecology, 262*(1-2), 79–90.

Hegazy, A. and Doust, J. L. (2016). *Plant Ecology in the Middle East*. Oxford University Press.

Heredia-Guerrero, J. A., Guzman-Puyol, S., Benítez, J. J., Athanassiou, A., Heredia, A. and Domínguez, E. (2018). Plant cuticle under global change: biophysical implications. *Global Change Biology, 24*(7), 2749–2751.

Hill, K. E., Guerin, G. R., Hill, R. S. and Watling, J. R. (2015). Temperature influences stomatal density and maximum potential water loss through stomata of Dodonaea viscosa subsp. angustissima along a latitude gradient in southern Australia. *Australian Journal of Botany, 62*(8), 657–665.

Holmes, M. G. and Keiller, D. R. (2002). Effects of pubescence and waxes on the reflectance of leaves in the ultraviolet and photosynthetic wavebands: a comparison of a range of species. *Plant Cell and Environment, 25*(1), 85–93. doi:DOI 10.1046/j.1365-3040.2002.00779.x.

Hoshino, R., Yoshida, Y. and Tsukaya, H. (2019). Multiple steps of leaf thickening during sun-leaf formation in Arabidopsis. *The Plant Journal, 100*(4), 738–753.

Hüner, N. P., Bode, R., Dahal, K., Busch, F. A., Possmayer, M., Szyszka, B., Rosso, D., Ensminger, I., Król, M., Ivanov, A. G. and Maxwell, D. (2013). Shedding some light on cold acclimation, cold adaptation, and phenotypic plasticity. *Botany, 91*(3), 127–136.

IUPS Thermal Commission, C. f. T. P. o. t. I. U. o. P. S. (1987). Glossary of terms for thermal physiology. *Pflügers Archiv-European Journal of Physiology, 410*, 567–587.

Johnson, M. P. and Wientjes, E. (2020). The relevance of dynamic thylakoid organisation to photosynthetic regulation. *Biochimica et Biophysica Acta (BBA)-Bioenergetics, 1861*(4), 148039.

Jordan, G. J. (1997). Uncertainty in palaeoclimatic reconstructions based on leaf physiognomy. *Australian Journal of Botany, 45*(3), 527–547.

Jumrani, K., Bhatia, V. S. and Pandey, G. P. (2017). Impact of elevated temperatures on specific leaf weight, stomatal density, photosynthesis and chlorophyll fluorescence in soybean. *Photosynthesis Research, 131*(3), 333–350.

Kadioglu, A., Terzi, R., Saruhan, N. and Saglam, A. (2012). Current advances in the investigation of leaf rolling caused by biotic and abiotic stress factors. *Plant Science, 182*, 42–48.

Kadochová, Š. and Frouz, J. (2014). Thermoregulation strategies in ants in comparison to other social insects, with a focus on red wood ants (Formica rufa group). *F1000Research, 2*(280), 1–15. doi:10.12688/f1000research.2-280.v2.

Kampowski, T., Demandt, S., Poppinga, S. and Speck, T. (2018). Kinematical, structural and mechanical adaptations to desiccation in poikilohydric Ramonda myconi (Gesneriaceae). *Frontiers in Plant Science, 9*, 1701.

Kasahara, M., Kagawa, T., Oikawa, K., Suetsugu, N., Miyao, M. and Wada, M. (2002). Chloroplast avoidance movement reduces photodamage in plants. *Nature, 420*(6917), 829–832.

Kerstiens, G. (1996). Signalling across the divide: a wider perspective of cuticular structure— function relationships. *Trends in Plant Science, 1*(4), 125–129.

Kevan, P. (1989). Thermoregulation in arctic insects and flowers: adaptation and co-adaptation in behaviour, anatomy, and physiology. *Thermal Physiology. Elsevier Science Publishers BV (Biomedical Division), Amsterdam*, 747–753.

Kevan, P. G. and Shorthouse, J. D. (1970). Behavioural thermoregulation by high arctic butterflies. *Arctic, 23*(4), 268–279.

Khazaei, H., Street, K., Santanen, A., Bari, A. and Stoddard, F. L. (2013). Do faba bean (Vicia faba L.) accessions from environments with contrasting seasonal moisture availabilities differ in stomatal characteristics and related traits? *Genetic Resources and Crop Evolution, 60*(8), 2343–2357.

Kirchhoff, H. (2014). Structural changes of the thylakoid membrane network induced by high light stress in plant chloroplasts. *Philosophical Transactions of the Royal Society B: Biological Sciences, 369*(1640), 20130225.

Knaust, D. (2012). Trace-fossil systematics. pp. 79–101. *In*: D. Knaust and R. G. Bromley (eds.). *Trace Fossils as Indicators of Sedimentary Environments* (Vol. 64). Amsterdam, Netherlands: Elsevier.

Konrad, W., Burkhardt, J., Ebner, M. and Roth-Nebelsick, A. (2015). Leaf pubescence as a possibility to increase water use efficiency by promoting condensation. *Ecohydrology, 8*(3), 480–492.

Korb, J. and Linsenmair, K. E. (2000). Ventilation of termite mounds: new results require a new model. *Behavioral Ecology, 11*(5), 486–494. doi:DOI 10.1093/beheco/11.5.486.

Lane, N. J. and Skaer, H. l. (1980). Intercellular junctions in insect tissues. pp. 35–214. *In*: Berridge, J. E., Treherne, J. E. and Wigglesworth, V. B. (eds.). *Advances in Insect Physiology* (Vol. 15). London, UK: Academic Press.

Lawes, M. J., Richards, A., Dathe, J. and Midgley, J. J. (2011). Bark thickness determines fire resistance of selected tree species from fire-prone tropical savanna in north Australia. *Plant Ecology, 212*(12), 2057–2069.

Legendre, L. J. and Davesne, D. (2020). The evolution of mechanisms involved in vertebrate endothermy. *Philosophical Transactions of the Royal Society B, 375*(1793), 20190136.

Leigh, A., Sevanto, S., Ball, M. C., Close, J. D., Ellsworth, D. S., Knight, C. A., Nicotra, A. B. and Vogel, S. (2012). Do thick leaves avoid thermal damage in critically low wind speeds? *New Phytologist, 194*(2), 477–487.

Leigh, A., Sevanto, S., Close, J. and Nicotra, A. (2017). The influence of leaf size and shape on leaf thermal dynamics: does theory hold up under natural conditions? *Plant, Cell & Environment, 40*(2), 237–248.

Lenssen, A., Banfield, J. and Cash, S. (2001). The influence of trichome density on the drying rate of alfalfa forage. *Grass and Forage Science, 56*(1), 1–9.

Leroy, C., Gril, E., Ouali, L. S., Coste, S., Gérard, B., Maillard, P., Merciere, H. and Stahl, C. (2019). Water and nutrient uptake capacity of leaf-absorbing trichomes vs. roots in epiphytic tank bromeliads. *Environmental and Experimental Botany, 163*, 112–123.

Li, Y., Wang, Z., Xu, X., Han, W., Wang, Q. and Zou, D. (2016). Leaf margin analysis of Chinese woody plants and the constraints on its application to palaeoclimatic reconstruction. *Global Ecology and Biogeography, 25*(12), 1401–1415.

López, R. P. and Valdivia, S. (2007). The importance of shrub cover for four cactus species differing in growth form in an Andean semi-desert. *Journal of Vegetation Science, 18*(2), 263–270.

Lunney, D., Crowther, M. S., Wallis, I., Foley, W. J., Lemon. J., Wheeler, R., Madani, G., Orscheg, C., Griffith, J. E., Krockenberger, M., Retamales, M. and Stalenberg, E. (2012). Koalas and climate change: a case study on the Liverpool Plains, north-west New South Wales. pp. 150–168. D. Lunney and P. Hutchings (eds.). *Wildlife and Climate Change: towards Robust Conservation Strategies for Australian Fauna.*

Mannuthy, T. (2017). Behavioral responses to livestock adaptation to heat stress challenges. *Asian Journal of Animal Sciences, 11*(1), 1–13.

Maricle, B. R., Cobos, D. R. and Campbell, C. S. (2007). Biophysical and morphological leaf adaptations to drought and salinity in salt marsh grasses. *Environmental and Experimental Botany, 60*(3), 458–467.

Martin, R. E., Asner, G. P., Bentley, L. P., Shenkin, A., Salinas, N., Huaypar, K. Q., Pillco, M. M., Ccori, A., Delis, F., Enquist, B. J., Diaz, S. and Malhi, Y. (2020). Covariance of sun and shade leaf traits along a tropical forest elevation gradient. *Frontiers in Plant Science, 10*, 1810.

Mauseth, J. D. (2000). Theoretical aspects of surface-to-volume ratios and water-storage capacities of succulent shoots. *American Journal of Botany, 87*(8), 1107–1115.

Mckee, J. and Richards, A. (1998). Effect of flower structure and flower colour on intrafloral warming and pollen germination and pollen-tube growth in winter flowering Crocus L. (Iridaceae). *Botanical Journal of the Linnean Society, 128*(4), 369–384.

Mershon, J. P., Becker, M. and Bickford, C. P. (2015). Linkage between trichome morphology and leaf optical properties in New Zealand alpine Pachycladon (Brassicaceae). *New Zealand Journal of Botany, 53*(3), 175–182.

Michaletz, S. T., Weiser, M. D., Zhou, J., Kaspari, M., Helliker, B. R. and Enquist, B. J. (2015). Plant thermoregulation: energetics, trait–environment interactions, and carbon economics. *Trends in Ecology & Evolution, 30*(12), 714–724.

Miguelsierra. (2011). Scheme Chloroplast-en. Retrieved from https://commons.wikimedia. org/wiki/File:Scheme_Chloroplast-en.svg, Creative Commons Attribution-Share Alike 4.0 International, 3.0 Unported, 2.5 Generic, 2.0 Generic and 1.0 Generic license: https://creativecommons.org/licenses/by-sa/4.0/, https://creativecommons.org/ licenses/by-sa/3.0/deed.en, https://creativecommons.org/licenses/by-sa/2.5/deed. en, https://creativecommons.org/licenses/by-sa/1.0/deed.en.

Milling, C. R., Rachlow, J. L., Chappell, M. A., Camp, M. J., Johnson, T. R., Shipley, L. A., Paul, D. R. and Forbey, J. S. (2018). Seasonal temperature acclimatization in a semi-fossorial mammal and the role of burrows as thermal refuges. *Peerj-The Journal of Life and Environmental Sciences, 6*, e4511. doi:10.7717/peerj.4511.

Monteith, J. L. and Mount, L. E. (eds.). (1973). *Heat Loss from Animals and Man: Assessment and Control.* London, UK: Butterworths.

Mu, J., Li, G. and Sun, S. (2010). Petal color, flower temperature, and behavior in an alpine annual herb, Gentiana leucomelaena (Gentianaceae). *Arctic, Antarctic, and Alpine Research, 42*(2), 219–226.

Needham, A. D., Dawson, T. J. and Hales, J. R. (1974). Forelimb blood flow and saliva spreading in the thermoregulation of the red kangaroo, Megaleia rufa. *Comparative Biochemistry and Physiology, 49*(3A), 555–565.

Newton, I. (2010). *The Migration Ecology of Birds.* London, UK: Elsevier.

Nicastro, K. R., Zardi, G. I., McQuaid, C. D., Pearson, G. A. and Serrão, E. A. (2012). Love thy neighbour: group properties of gaping behaviour in mussel aggregations. *PLoS One, 7*(10), e47382.

Nielsen, M. E. and Papaj, D. R. (2015). Effects of developmental change in body size on ectotherm body temperature and behavioral thermoregulation: caterpillars in a heat stressed environment. *Oecologia, 177,* 171–179.

Niinemets, U. (2007). Photosynthesis and resource distribution through plant canopies. *Plant, Cell & Environment, 30*(9), 1052–1071.

Niinemets, U. (2010). A review of light interception in plant stands from leaf to canopy in different plant functional types and in species with varying shade tolerance. *Ecological Research, 25*(4), 693–714.

Ocko, S. A. and Mahadevan, L. (2014). Collective thermoregulation in bee clusters. *Journal of the Royal Society Interface, 11*(91), 1–14. doi:10.1098/rsif.2013.1033.

Paladino, F. V., O'Connor, M. P. and Spotila, J. R. (1990). Metabolism of leatherback turtles, gigantothermy, and thermoregulation of dinosaurs. *Nature, 344*(6269), 858–860.

Patel, D. and Franklin, K. A. (2009). Temperature-regulation of plant architecture. *Plant Signaling & Behavior, 4*(7), 577–579.

Penacchio, O., Cuthill, I. C., Lovell, P. G., Ruxton, G. D. and Harris, J. M. (2015). Orientation to the sun by animals and its interaction with crypsis. *Functional Ecology, 29*(9), 1165–1177. doi:10.1111/1365-2435.12481.

Ratnakaran, A. P., Sejian, V., Sanjo Jose, V., Vaswani, S. and Bagath, M. (2017). Behavioral responses to livestock adaptation to heat stress challenges. *Asian Journal of Animal Sciences, 11*(1), 1–13.

Reyes-Olivas, A., García-Moya, E. and López–Mata, L. (2002). Cacti–shrub interactions in the coastal desert of northern Sinaloa, Mexico. *Journal of Arid Environments, 52*(4), 431–445.

Richards, S.-A. (1970). Physiology of thermal panting in birds. *Annales de Biologie Animale Biochimie Biophysique, 10*(Hors-série 2), 151–168.

Riederer, M. and Schreiber, L. (2001). Protecting against water loss: analysis of the barrier properties of plant cuticles. *Journal of Experimental Botany, 52*(363), 2023–2032.

Riederer, M. (2018). *Introduction: Biology of the Plant Cuticle.* Annual Plant Review Series.: Volume 23. Wiley on-line.

Robson, T. M., Klem, K., Urban, O. and Jansen, M. A. (2015). Re-interpreting plant morphological responses to UV-B radiation. *Plant, Cell & Environment, 38*(5), 856–866.

Rodriguez, A., Derita, M. G., Borkosky, S. A., Socolsky, C., Bardón, A. and Hernández, M. (2018). Bioactive farina of Notholaena sulphurea (Pteridaceae): Morphology and histochemistry of glandular trichomes. *Flora, 240,* 144–151.

Rowe, M., Bakken, G., Ratliff, J. and Langman, V. (2013). Heat storage in Asian elephants during submaximal exercise: behavioral regulation of thermoregulatory constraints on activity in endothermic gigantotherms. *Journal of Experimental Biology, 216*(10), 1774–1785.

Ryeland, J., Weston, M. A. and Symonds, M. R. (2017). Bill size mediates behavioural thermoregulation in birds. *Functional Ecology, 31*(4), 885–893.

Sanusi, R. and Livesley, S. J. (2020). London Plane trees (Platanus x acerifolia) before, during and after a heatwave: Losing leaves means less cooling benefit. *Urban Forestry & Urban Greening,* 126746.

Sargent, F. (1961). The mechanism of hidromeiosis. *International Journal of Biometeorology, 5*(1), 37–40.

Schaaf, A. A., Garcia, C. G. and Greeney, H. F. (2020). Nest orientation in closed nests of Passeriformes across a latitudinal gradient in the Southern Neotropic. *Acta Ornithologica, 54*(2), 263–268.

Schonherr, J. and Merida, T. (1981). Water permeability of plant cuticular membranes: the effects of humidity and temperature on the permeability of non-isolated cuticles of onion bulb scales. *Plant, Cell & Environment, 4*(5), 349–354.

Schuster, A.-C., Burghardt, M., Alfarhan, A., Bueno, A., Hedrich, R., Leide, J., Thomas, J. and Riederer, M. (2016). Effectiveness of cuticular transpiration barriers in a desert plant at controlling water loss at high temperatures. *AoB Plants, 8*(1), plw027.

Schymanski, S. J., Or, D. and Zwieniecki, M. (2013). Stomatal control and leaf thermal and hydraulic capacitances under rapid environmental fluctuations. *PLoS One, 8*(1), e54231.

Secor, S. M., Jayne, B. C. and Bennett, A. F. (1992). Locomotor performance and energetic cost of sidewinding by the snake Crotalus cerastes. *Journal of Experimental Biology, 163*(1), 1–14.

Seeley, T. D. (2010). *Honeybee Democracy*. Princeton, NJ: Princeton University Press.

Seigel, R. A., Collins, J. T. and Novak, S. S. (1987). *Snakes: Ecology and Evolutionary Biology*. New York, NY: Macmillan.

Sekiya, N. and Yano, K. (2008). Stomatal density of cowpea correlates with carbon isotope discrimination in different phosphorus, water and CO_2 environments. *New Phytologist, 179*(3), 799–807.

Seuront, L., Ng, T. P. and Lathlean, J. A. (2018). A review of the thermal biology and ecology of molluscs, and of the use of infrared thermography in molluscan research. *Journal of Molluscan Studies, 84*(3), 203–232.

Shea, N. (2019). Inside the harsh lives of wolves living at the top of the world. *National Geographic*, p. 116.

Shelef, O., Summerfield, L., Lev-Yadun, S., Villamarin-Cortez, S., Sadeh, R., Herrmann, I. and Rachmilevitch, S. (2019). Thermal benefits from white variegation of Silybum marianum leaves. *Frontiers in Plant Science, 10*, 688.

Shibasaki, M. and Crandall, C. G. (2010). Mechanisms and controllers of eccrine sweating in humans. *Frontiers in Bioscience (Scholar Edition), 2*, 685.

Shinozaki, K. and Yamaguchi-Shinozaki, K. (2000). Molecular responses to dehydration and low temperature: differences and cross-talk between two stress signaling pathways. *Current Opinion in Plant Biology, 3*(3), 217–223.

Shrestha, M., Garcia, J. E., Bukovac, Z., Dorin, A. and Dyer, A. G. (2018). Pollination in a new climate: assessing the potential influence of flower temperature variation on insect pollinator behaviour. *PLoS One, 13*(8), e0200549.

Sinha, A., Misra, J. C. and Shit, G. C. (2016). Effect of heat transfer on unsteady MHD flow of blood in a permeable vessel in the presence of non-uniform heat source. *Alexandria Engineering Journal, 55*(3), 2023–2033. doi:10.1016/j.aej.2016.07.010.

Smith, S. D., Monson, R. and Anderson, J. E. (2012). *Physiological Ecology of North American Desert Plants*. Springer Science & Business Media.

Solbrig, O. T. and Orians, G. H. (1977). The adaptive characteristics of desert plants: a cost/benefit analysis of photosynthesis leads to predictions about the types and distributions of desert plants. *American Scientist, 65*(4), 412–421.

Song, Y., Chen, Q., Ci, D., Shao, X. and Zhang, D. (2014). Effects of high temperature on photosynthesis and related gene expression in poplar. *BMC Plant Biology, 14*(1), 111.

Stuart-Fox, D., Newton, E. and Clusella-Trullas, S. (2017). Thermal consequences of colour and near-infrared reflectance. *Philosophical Transactions of the Royal Society B, Biological Science, 372*(1724), 20160345. doi:10.1098/rstb.2016.0345.

Suetsugu, N. and Wada, M. (2020). Signalling mechanism of phototropin-mediated chloroplast movement in Arabidopsis. *Journal of Plant Biochemistry and Biotechnology*, 1–10.

Sultan, S. E. (2015). *Organism and Environment: Ecological Development, Niche Construction, and Adaptation*. New York, NY: Oxford University Press.

Terashima, I., Araya, T., Miyazawa, S.-I., Sone, K. and Yano, S. (2005). Construction and maintenance of the optimal photosynthetic systems of the leaf, herbaceous plant and tree: an eco-developmental treatise. *Annals of Botany, 95*(3), 507–519.

Terrien, J., Perret, M. and Aujard, F. (2011). Behavioral thermoregulation in mammals: a review. *Frontiers in Bioscience, 16*, 1428–1444.

Thomas, B., Murphy, D. J. and Murray, B. G. (2016). *Encyclopedia of Applied Plant Sciences*. Academic Press.

Thomas, D. N. and Fogg, G. E. (2008). *The Biology of Polar Regions*. New York, NY: Oxford University Press.

Trochet, A., Dupoué, A., Souchet, J., Bertrand, R., Deluen, M., Murarasu, S., Calvez, O., Martinez-Silvestre, A., Verdaguer-Foz, I., Darnet, E., Chevalier, H. L., Mossoll-Torres, M., Guillaume, O. and Aubret, F. (2018). Variation of preferred body temperatures along an altitudinal gradient: A multi-species study. *Journal of Thermal Biology, 77*, 38–44.

Valera, F., Díaz-Paniagua, C., Garrido-García, J. A., Manrique, J., Pleguezuelos, J. M. and Suárez, F. (2011). History and adaptation stories of the vertebrate fauna of southern Spain's semi-arid habitats. *Journal of Arid Environments, 75*(12), 1342–1351.

van der Kooi, C. J., Kevan, P. G. and Koski, M. H. (2019). The thermal ecology of flowers. *Annals of Botany, 124*(3), 343–353.

van Doorn, W. G. and van Meeteren, U. (2003). Flower opening and closure: a review. *Journal of Experimental Botany, 54*(389), 1801–1812.

Vaughan, T. A., Ryan, J. M. and Czaplewski, N. J. (2013). *Mammalogy*. Sudbury, MA: Jones & Bartlett Publishers.

Venturas, M. D., Sperry, J. S. and Hacke, U. G. (2017). Plant xylem hydraulics: what we understand, current research, and future challenges. *Journal of Integrative Plant Biology, 59*(6), 356–389.

Vicenzi, N., Ibargüengoytía, N. and Corbalan, V. (2019). Activity patterns and thermoregulatory behavior of the viviparous lizard Phymaturus palluma in Aconcagua Provincial Park, Argentine Andes. *Herpetological Conservation and Biology, 14*(2), 337–348.

Vogelmann, T. and Martin, G. (1993). The functional significance of palisade tissue: penetration of directional versus diffuse light. *Plant, Cell & Environment, 16*(1), 65–72.

Wagner, G. J. (1991). Secreting glandular trichomes: more than just hairs. *Plant Physiology, 96*(3), 675–679.

Wahid, A., Gelani, S., Ashraf, M. and Foolad, M. R. (2007). Heat tolerance in plants: an overview. *Environmental and Experimental Botany, 61*(3), 199–223.

Walsberg, G. E. (1988). The significance of fur structure for solar heat gain in the rock squirrel, Spermophilus variegatus. *Journal of Experimental Biology, 138*(1), 243–257.

Wang, Z., Xu, J-h., Mou, J-j., Kong, X-t., Zou, J-w., Xue, H-l., Wu, M. and Xu, L-x. (2020). Novel ultrastructural findings on cardiac mitochondria of huddling Brandt's voles in mild cold environment. *Comparative Biochemistry and Physiology Part A: Molecular & Integrative Physiology, 249*, 110766.

Wani, S. H. and Kumar, V. (2020). *Heat Stress Tolerance in Plants: Physiological, Molecular and Genetic Perspectives*. John Wiley & Sons.

Waraich, E. A., Ahmad, R., Halim, A. and Aziz, T. (2012). Alleviation of temperature stress by nutrient management in crop plants: a review. *Journal of Soil Science and Plant Nutrition, 12*(2), 221–244. doi:10.4067/S0718-95162012000200003.

Watling, J. R., Grant, N. M., Miller, R. E. and Robinson, S. A. (2008). Mechanisms of thermoregulation in plants. *Plant Signaling & Behavior, 3*(8), 595–597.

Watson, G. S., Gregory, E. A., Johnstone, C., Berlino, M., Green, D. W., Peterson, N. R., Schoeman, D. S. and Watson, J. A. (2018). Like night and day: Reversals of thermal gradients across ghost crab burrows and their implications for thermal ecology. *Estuarine, Coastal and Shelf Science, 203*, 127–136.

Wharton, D. A. and Brown, I. M. (1991). Cold-tolerance mechanisms of the Antarctic Nematode Panagrolaimus-Davidi. *Journal of Experimental Biology, 155*(1), 629–641.

Whittaker, M. and Thomas, V. (1983). Seasonal levels of fat and protein reserves of snowshoe hares in Ontario. *Canadian Journal of Zoology, 61*(6), 1339–1345.

Wickens, G. E. (2013). *Ecophysiology of Economic Plants in Arid and Semi-Arid Lands.* Springer Science & Business Media.

Wilms, T. M., Wagner, P., Shobrak, M., Rödder, D. and Böhme, W. (2011). Living on the edge?—On the thermobiology and activity pattern of the large herbivorous desert lizard Uromastyx aegyptia microlepis Blanford, 1875 at Mahazat as-Sayd Protected Area, Saudi Arabia. *Journal of Arid Environments, 75*(7), 636–647. doi:10.1016/j.jaridenv.2011.02.003.

Wisniewski, M., Gusta, L. and Neuner, G. (2014). Adaptive mechanisms of freeze avoidance in plants: a brief update. *Environmental and Experimental Botany, 99*, 133–140.

Withers, P. C. and Campbell, J. D. (1985). Effects of environmental cost on thermoregulation in the desert Iguana. *Physiological Zoology, 58*(3), 329–339. doi:DOI 10.1086/physzool.58.3.30156004.

Wright, I. J., Dong, N., Maire, V., Prentice, I. C., Westoby, M., Díaz, S., Gallagher, R. V., Jacobs, B. F., Kooyman, R., Law, E. A., Leishman, M. R., Niinemets, Ü., Reich, P. B., Sack, L., Villar, R., Wang, H. and Wilf, P. (2017). Global climatic drivers of leaf size. *Science, 357*(6354), 917–921.

Wu, G., Liu, H., Hua, L., Luo, Q., Lin, Y., He, P., Feng, S., Liu, J. and Ye, Q. (2018). Differential responses of stomata and photosynthesis to elevated temperature in two co-occurring subtropical forest tree species. *Frontiers in Plant Science, 9*, 467.

Wu, N. C., Alton, L. A., Clemente, C. J., Kearney, M. R. and White, C. R. (2015). Morphology and burrowing energetics of semi-fossorial skinks (Liopholis spp.). *Journal of Experimental Biology, 218*(15), 2416–2426.

Xiao, X., Gu, Y., Wu, G., Zhang, D. and Ke, H. (2019). Controllable crimpness of animal hairs via water-stimulated shape fixation for regulation of thermal insulation. *Polymers, 11*(1), 172.

Xiao, Y., Tholen, D. and Zhu, X.-G. (2016). The influence of leaf anatomy on the internal light environment and photosynthetic electron transport rate: exploration with a new leaf ray tracing model. *Journal of Experimental Botany, 67*(21), 6021–6035.

Yorzinski, J. L., Lam, J., Schultz, R. and Davis, M. (2018). Thermoregulatory postures limit antipredator responses in peafowl. *Biology Open, 7*(1).

Yoshioka-Nishimura, M. (2016). Close relationships between the PSII repair cycle and thylakoid membrane dynamics. *Plant and Cell Physiology, 57*(6), 1115–1122.

Young, S. R. and Block, W. (1980). Experimental studies on the cold tolerance of Alaskozetes-Antarcticus. *Journal of Insect Physiology, 26*(3), 189–200. doi:Doi 10.1016/0022-1910(80)90080-3.

Zephyris. (2011). Leaf Tissue Structure. Retrieved from https://commons.wikimedia.org/wiki/File:Leaf_Tissue_Structure.svg, Creative Commons Attribution-Share Alike 3.0 Unported: https://creativecommons.org/licenses/by-sa/3.0/deed.en.

Zhu, J., Yu, Q., Xu, C., Li, J. and Qin, G. (2018). Rapid estimation of stomatal density and stomatal area of plant leaves based on object-oriented classification and its ecological trade-off strategy analysis. *Forests, 9*(10), 616.

Chapter 7
Parallels in Building Design

7.1 Introduction

Having categorised the thermal adaptation mechanisms of organisms in Chapter 6, this chapter introduces a number of potential parallels in building design. The aim is to explore how many are already used and whether there are strategies which could inspire designers to improve the thermal performance and hence, the energy efficiency of buildings.

As in Chapter 6, any parallel building design mechanism is categorized as either passive or active. To enable easy comparison of equivalent strategies, the biological mechanisms used by both animals and plants and their parallels in building design are placed alongside each other with the energy efficiency design parallels on the right-hand sides of Figures 7-1, 7-3, and 7-4 This step is where the initial structure of the ThBA took shape.

As discussed in Chapter 4, the ThBA was developed to connect thermal challenges found in buildings to relevant biological strategies. Therefore, it was essential to understand how biological mechanisms occurred in nature in order to evaluate their potential to be used in building design. Understanding the thermoregulatory mechanisms used by organisms also revealed whether these mechanisms have been used or do have the potential to be used in energy-efficient building design as either passive or active design strategies.

The left-hand side of Figures 7-1, 7-3 and 7-4 show biological thermal adaptation strategies while the equivalent building design strategies are shown on the right-hand side of the figure. The grey column in the centre outlines the significant functional and morphological principals for each individual strategy. In the context of bio-inspired design (BID), these principles need to be extracted independently from the biological systems to facilitate the development of design solutions.

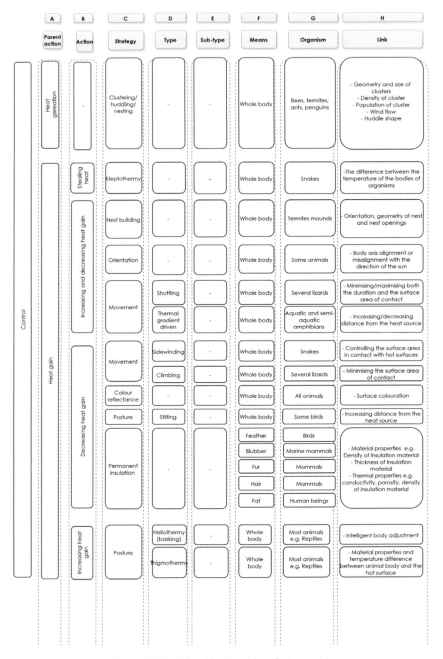

Figure 7-1, Part One (Continued on facing page)

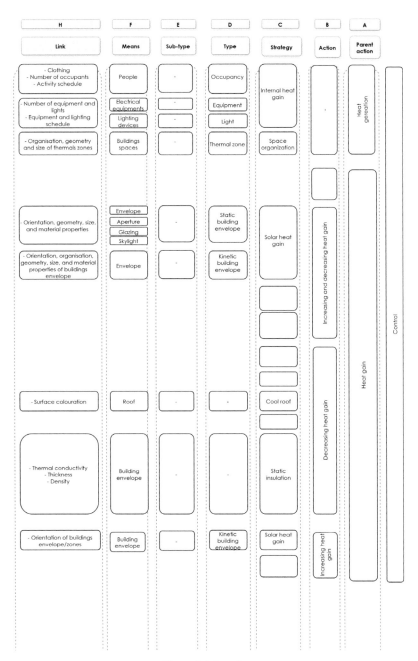

Figure 7-1, Part One

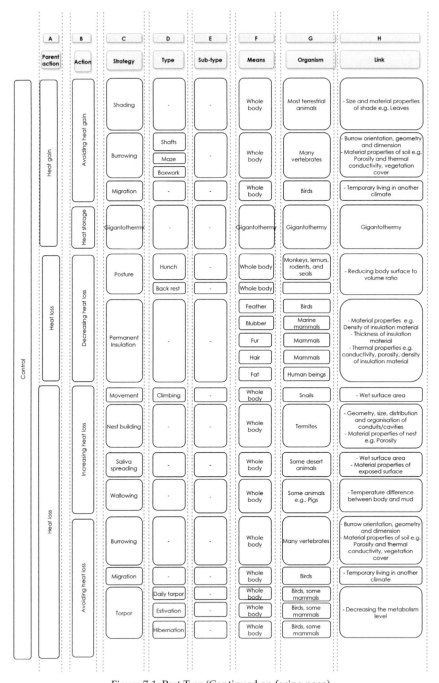

Figure 7-1, Part Two (Continued on facing page)

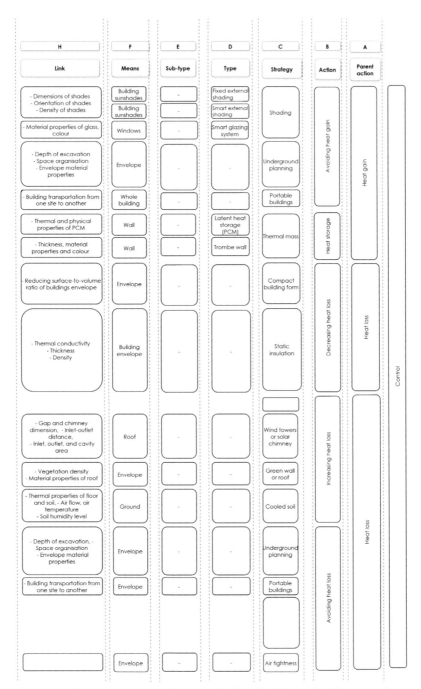

Figure 7-1. Passive strategies used by animals aligned with their building equivalent.

The list of thermoregulatory mechanisms for buildings suggest that energy-efficient building design has already benefited from imitating many thermal adaptation mechanisms that nature employs for survival, except for the areas shown as white cells. These represent groups of biological thermal adaptation mechanisms yet to be investigated or evaluated by architects. Although there may have been prior mention of possible design implications for some mechanisms, no comprehensive practical approach has been taken to investigate their usefulness in building design. An example would be a light-collecting system based on lens-shaped epidermal cells (see 6.4.2). This system could be developed if the surface area of a building was reduced to save energy as a means of bringing light into the building interior, but this would be an example of combining thermoregulatory strategies.

The design implications in the context of this book exclude the morphological limitation of biological strategies along with the one-off suggestions for which their application in buildings has not gone beyond either prototypes or models produced by simulation software. This means these empty white cells present challenges to designers.

7.2 Passive methods of thermal regulation in buildings

This section introduces passive design strategies that architects use to improve energy efficacy in buildings. These are parallel to behavioural thermoregulatory strategies for animals.

7.2.1 Generating heat

Interior heat gain: In buildings, internal heat gain can occur through heat generated by occupants, equipment and lighting. Occupancy can be regarded as a parallel to clustering, as people can move and cluster within buildings. For example, in some places, people traditionally gather in one heated room during winter rather than spreading out through the whole building. In northern medieval monasteries, only the calefactory was heated and therefore became the part of the building in which people clustered.

Space organisation: Clustering may also suggest space-heating strategies as the heat generated by clusters from their movement and amalgamation patterns can be paralleled by HVAC systems moving heat around a building. As discussed earlier, the aggregation of animals in herds, flocks and swarms is done to optimise heat generation and has inspired engineering design. An example of this is the development of an evolutionary computation technique for designing heat sinks based on swarm optimisation (Alrasheed 2011). In the context of swarm intelligence

(SI), adaptive and decentralised HVAC operation management could be informed by the self-grouping of animals, and in this way, efficiently control heating and cooling systems.

Stigmergy is an action which leaves a trace, leading to another action. Although it is a cognitive behaviour in insects, it is similar to SI (a reactive behaviour). Stigmergy has inspired major data management algorithms which are named after insect behaviours (Awad and Khanna 2015). Examples include artificial immune system(s), artificial bee colony, bacterial foraging optimisation, ant colony optimisation and particle swarm.

The collective behaviour of insects in response to high and low temperatures, such as the clustering of bees and movement patterns in hives and when swarming, might potentially be emulated in space arrangements in architectural design (Ocko and Mahadevan 2014). The energy efficiency of buildings could be improved by translating the parameters that affect heat transfer in a bee swarm. For instance, the density of bees in different parts of a swarm could inspire HVAC distribution or space planning. To achieve this, a mathematical heat transfer equation of the movement of bees within a cluster could be a helpful addition to building energy modelling (Figure 7-2). Ocko and Mahadevan (2014) introduced the following variables for the calculation of heat generation based on bee clusters within a spherical boundary:

- Schematic of interior (Ω)
- ($\delta\Omega$) boundary mantle–core structure
- $\hat{S}$ and $\check{Z}$ are radial, vertical directions in polar coordinates
- R is the cluster radius

Similarly, building plans and HVAC distribution could be inspired by the collective structures of ant armies. For example, Anderson et al. (2002) found at least four different self-assemblage strategies made by ants that affect thermoregulation (Table 7-1).

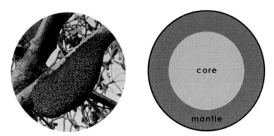

Figure 7-2. Heat generation in a bee cluster (Left) the swarm on a branch; (Right) theoretical section through a swarm (Author, adapted from Ocko and Mahadevan (2014) and Balouria (2019)).

Table 7-1. Social insect self-assemblages classified by structure (Anderson et al. 2002).

Self-assemblage	Structure		
	Chain (1D)	Mesh (2D)	Cluster (3D)
Bivouac	*	*	*
Curtains		*	
Swarms			*
Thermoregulatory			*

7.2.2 Controlling heat gain

Increasing and decreasing heat gain

Solar heat gain

1) *Static building envelope*: There are obvious similarities between building envelopes and nest building in organisms because the solar radiation received by both is influenced by orientation. The size and shape of a building could be seen as being similar to the body of animals as the level of heat gain can be controlled by adjusting these factors. At a smaller scale, glazing systems and apertures such as skylights in the building envelope could be seen as conceivable parallels of nests in building design, as the orientation, shape and size, material and type of these openings in the building envelope will affect heat gain and loss. The historic use of small apertures in hot, desert climates is an example of this, something that is ignored with the modern all-glass building in a hot climate, which has to be actively controlled using a HVAC system to avoid overheating.

   As discussed earlier (see nest building, 6.1.2), the nest entrance plays a key role in solar heat gain. While organisms adjust the shape, size and the slope of the entrance to their nests, architects alter the shape, size and angles of apertures in the building envelop to optimise solar heat gain in climates where this is the goal. The entrance to the houses of the Inuit is a tunnel arranged so as to avoid the entry of cold winds (Whitridge 2008). Given this, the thoughtful placing of openings in the building envelope can be considered to have a parallel in nest building.

2) *Kinetic building envelope*: The Heliotrope House in Freiburg, which was designed to follow the sun, is an example of the kinetic approach to thermal design (Ramzy and Fayed 2011). This has a direct parallel with heliotropism in plants such as the sunflower which over the day turns the flower head to follow the sun. Even in non-moving buildings, movable blinds and shutters could be viewed as kinetic elements, as moving these can help to control solar gain. The opening of windows

to allow for exchange of gases can also be seen as having a parallel with the stomata of leaves. The latter open to allow the release of water vapour and the intake of oxygen and carbon dioxide. Similarly, windows are opened to allow for the release of water vapour and the intake of fresh air containing oxygen, the difference being that windows allow for the release of carbon dioxide which can build up from the exhalations of the occupants in a poorly ventilated building.

The similarity between moving architecture and organisms may not seem obvious as most buildings are static and generally almost all of their elements are in a fixed position. This contrasts with the dynamic ability of animals to move to a more appropriate microclimate to control heat gain. Comparing the number of kinetic to fixed buildings suggests that kinetic architecture may not be a useful approach (Fox 2016), although many efforts have been taken to incorporate movement into design. However, such movement takes energy. The latter has been sourced by integrating a number of sources, such as human movement, environmental kinetic forces (wind, air, water) and machines which use mechanical motors (El Razaz 2010). However, it is essential to take into account the costs and benefits of these approaches, which is similar to what organisms do in their energy trade-offs.

Decreasing heat gain

Cool roof: The colour reflectance of the skin and coats of animals seems to have a similar effect to a cool roof designed to reduce solar absorption (6.1.2, colour reflectance). To reduce solar heat gain, such a roof needs to be made of a material which has either a high solar reflectance or solar emittance (Roberts 2008). The roof energy balance of a building is affected by the overall thermal resistance of the roof material, together with the temperature difference between the inner and outer surface of the roof (Nikolaou et al. 2015), as shown in the following equation:

- $Q_{in} = (T_s - T_c)/R$ where:

  R is Overall thermal resistance of the roof material (m^2K/W),
  T_s is Temperature of the outer surface of the roof (K),
  And T_c is Temperature of the inner surface of the roof (K).

A study of a cool roof in central California found there was a 15% annual energy saving and a 20% annual energy cost saving (Rosado et al. 2014).

Static thermal insulation: Thermal insulation technology falls into the four major groups of bulk technology, reflective technology, nanotechnology and vacuum technology (Lee et al. 2016). Irrespective of their specific types, thermal conductivity, thickness and density of materials affect the effectiveness of each insulation technology. Bulk insulation has a history as long as building, with the thatched roof, whether of reeds, grasses or heather providing both insulation and a means of keeping the rain out of the building. Reflective insulation materials include aluminium bubble foil sheets. This is used as a roofing material for sarking to reflect solar heat gain in the summer, thus acting like a white-painted 'cool' roof. A traditional use of a reflective layer for insulation is found in the snow igloo. Once constructed, the heat inside from a small lamp and body heat will melt the internal surface which then freezes because of its immediate contact with the snow blocks, thus forming a reflective ice layer (González-Espada et al. 2001). Vacuum double glazing, where there is a vacuum between the two panes of glass rather than an inert gas like argon, has been under development since the 1990s, although the technology has yet to make serious inroads into the market place (Kocer 2020). A window consortium claims it is still under development (Efficient Windows Collaborative 2021). Vacuum-insulated panels where a vacuum is introduced into "…an encapsulated microporous material" have also been the subject of experiments, although the panels are fragile and deteriorate with time (Baetens et al. 2010). The application of nanotechnology to buildings is also still in the experimental stage (Torgal 2016). Of interest for this book is that, with the exceptions of bulk and reflective insulation materials, vacuum and nano materials have no parallels in the earth-bound natural world, rather nanobiotechnology is the application of nanotechnology to the fields of biology and medicine.

Avoiding heat gain

Shading: While obvious, the shading strategy in nature is equivalent to the design of solar shading systems for buildings. Many studies have analysed the effectiveness of shading systems on the cooling energy requirements of buildings (Tzempelikos and Athienitis 2007, Pino et al. 2012, Bellia et al. 2013). Shades can be either fixed or moveable. Fixed solar shading can be composed of single or multiple blades orientated vertically or horizontally. For example, in a traditional approach, people use whitewash to shade their glass houses in the summer months, removing this in the other seasons. The traditional plantation shutter as found on homes in the southern USA could be closed across the window to reduce solar gain, so these shading devices are also a kinetic part of the building envelope. Similar shutters are found in the houses of those living in hot, humid climates so that the openings can be shaded while allowing any cooling breezes to pass through the house.

Underground architecture: This corresponds to a burrowing strategy in which an animal seeks a favourable microclimate to avoid harsh environments (Yang et al. 2014). For example, some traditional Iranian houses have underground rooms, which were used on the hottest days (Beigli and Lenci 2016).

Portable building: Animal and bird migration links to the idea of portable buildings but most nomadic cultural movement is based on finding pastures rather than migrating to find a more suitable microclimate. In the European Alps, people would move to live in huts high in the mountains in summer so they could look after their flocks, which also migrated up the mountain to take advantage of the new grass growth.

Heat storage

Thermal mass: In architecture, the use of thermal mass is a strategy of both traditional and modern passive solar design. The former is found in hot, desert environments where the nights are cold. The construction material of the walls and roofs heats up during the day and as the temperature drops heat flows both into the building and out to the night air, thus cooling the construction material, which is ready to be heated up again the next day. In modern passive solar design for cold climates, the thermal mass has to be insulated from the outside so that the heat entering it from the sun is stored in the mass for days when the sun is not shining. Thermal storage can be achieved through either conventional mass materials like brick or concrete or phase change materials (Khudhair and Farid 2004). Phase change materials as discussed later (Section 7.3.2).

7.2.3 Controlling heat loss

Decreasing heat loss

Compact building form: One of the functions of building morphology is to control heat loss. Ratti et al. (2005) used energy analysis to establish the most efficient surface-to-volume ratio for building geometry. While such studies explore the optimised geometrical configurations for decreasing heat loss through the building envelope, there is a limit to using this approach, as a reduction in surface area limits opportunities for providing daylight and ventilation for internal spaces. The traditional hemispherical igloo again provides an example of a compact building form, although the form also relates to how the shaped ice blocks were assembled in a spiral form (Smith 2017); however, an igloo has no windows and limited ventilation. An animal reducing its surface-to-volume ratio by crouching has no such issues.

Static thermal insulation: See 7.2.2

Increasing heat loss

Solar chimney (wind tower): A solar chimney or wind tower is a form of passive cooling in buildings, the function of which is similar to that of termite mounds, as the structure of both enhances ventilation. The three components of a solar chimney are an inlet to capture the wind, a cooling cavity where the wind passes over water in a cistern and an outlet that allows the cooled air to enter the room (Aboulnaga 1998). The length and width of the chimney, the air gap dimensions, the inlet-outlet distance and area of the inlet, outlet and the cavity all affect its performance (Aboulnaga 1998). The solar chimney is a traditional feature of houses in Yadz, Iran, which has a desert climate (Niglio 2018). Wind towers have been modernised through use of "…integrated advanced building principles and technology" (Khan et al. 2008).

Green roof/wall: The incorporation of green roofs into building design can improve heat transfer through the building envelope. Green roofs can help to improve the microclimate, minimize the urban heat island effect and reduce energy consumption both at the building and urban scales (Morakinyo et al. 2017). Like saliva spreading, which kangaroos use as a cooling mechanism when at rest (Dawson 1974), the evaporation from a green roof could help in keeping a building cool. Green roofs generally have insulation, waterproofing and vegetation layers, depending on the building requirements in a particular location. Apart from having an equivalent in nature, the green roof has been described as providing "…ecosystem services in urban areas, including improved storm-water management, better regulation of building temperatures, reduced urban heat-island effects, and increased urban wildlife habitat" (Oberndorfer et al. 2007). A weak parallel from the natural world are the hermit crabs that can place sea anemones on their shells, though this may have more to do with camouflage than thermoregulation (Gusmão et al. 2020).

Cooled soil: The cooled soil strategy is to some extent similar to wallowing, as the building is raised over constantly moist soil. This has the potential to function as a cooling source during the summer in arid regions (Givoni 2007).

7.3 Active methods of thermal regulation in buildings

In this section, energy-efficient building design strategies have been characterised as analogous to the physiological thermal adaptation mechanisms performed by animals. Similar to Figure 7-1, Figure 7-3 follows the same structure to show the comparability of active biological solutions to similar strategies in architectural design.

7.3.1 Generating heat

Non-renewable resources: The operation of HVAC systems manually controlled by building occupants seems equivalent to non-shivering thermogenesis taking place in brown fat, as the energy used for running the HVAC equipment is not necessarily renewable. The difference is that in the body of animals, the mechanism of non-shivering heat generation takes place involuntarily while the manual control of HVAC systems is voluntary. Perhaps the intelligent control of HVAC systems could be seen as a more suitable parallel to involuntary non-shivering thermogenesis in animals (Parameshwaran et al. 2012).

Renewable resources (weak parallel): In shivering thermogenesis, the shaking muscles are responsible for generating heat. This is an iterative mechanism which can happen again and again, making it similar to heat generation using renewable sources of energy, such as the sun and wind. Heat generation in shivering processes can be produced either voluntarily or involuntarily. The latter takes place due to the cold-induced activation of muscle contractions while the former occurs during exercise. However, this could be a seen as a superficial comparison, and a more precise understanding of these two mechanisms could affect the validity of the analogy.

PV solar systems and wind turbines: Hybrid photovoltaic thermal solar systems (PV/T or PVT) consist of PV modules that convert solar radiation into electricity while also being connected to water or air heat extraction devices, since electricity production falls as the PV panel gets warmer. The heat energy exhausted from the PV module is transferred to internal spaces through water or air. In wind turbines, the rotating energy is converted into electrical energy. The analogy between non-shivering thermogenesis as a physiological thermal adaptation strategy and both PV solar systems and wind turbines is weak.

7.3.2 Controlling heat gain

Increasing heat gain

Ground cooling and heating: One of the strategies for controlling heat gain and heat loss in buildings is to use a heat exchanger system buried in the ground. Such a system is parallel to the countercurrent heat exchange mechanism in animals as it is composed of a circulatory network of pipes that use air or water as the heat transfer medium (Florides and Kalogirou 2007). In winter, the soil acts as a heat source while it performs a heat sink function in summer. Such a heat exchange system can be either open

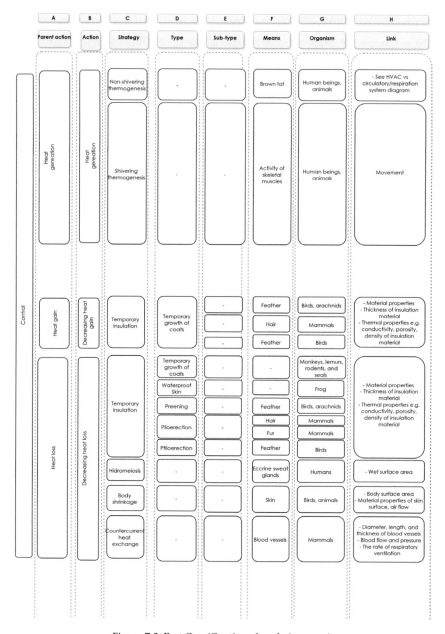

Figure 7-3, Part One (Continued on facing page)

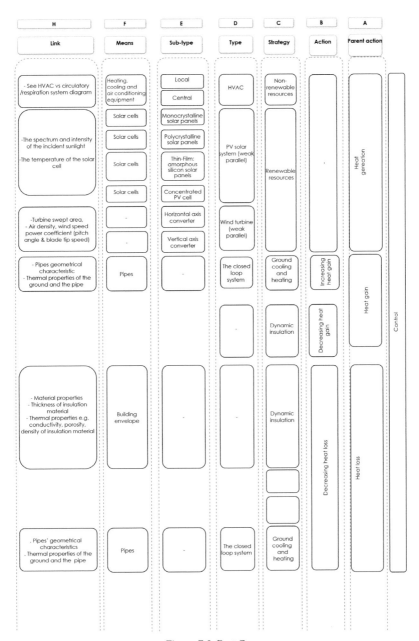

Figure 7-3, Part One

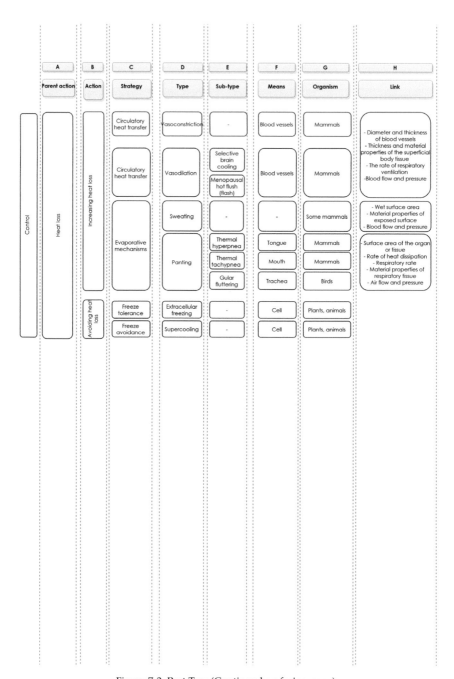

Figure 7-3, Part Two (Continued on facing page)

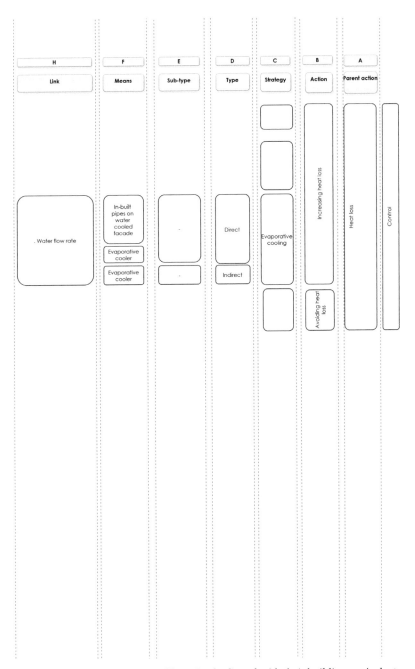

Figure 7-3. Active strategies used by animals aligned with their building equivalent.

or closed. In the latter, antifreeze is circulated through a closed loop in the ground and this passes through a heat exchanger, which transfers the energy to the refrigerant in the heat pump. In the open system, the refrigerant is pumped through the coils in the ground. The closed loop system is thus most like the countercurrent heat exchange in animals and birds.

The heat balance calculation of an underground heat exchange system uses the following input data: the geometrical characteristics of the system, the thermal characteristics of the ground and the pipes, and the undisturbed ground temperature during the operation of the system. These parameters seem analogous to the radius of blood vessels and their temperature, and the thermal properties of skin tissue.

Decreasing heat gain

PCM kinetic thermal insulation: Using a phase change material (PCM) as a means of insulation in walls and ceiling seems analogous to the 'temporary insulation' strategies of impermanent growth of coats, piloerection and ptiloerection used by animals. PCMs undergo a daily melting/solidifying cycle to reduce cooling and heating loads in buildings. Energy goes into the PCM to melt it and is emitted as heat during the solidifying phase. When cooling is important, PCMs can reduce the maximum cooling load and hence, the size of the HVAC equipment by smoothing out the demand (Halford and Boehm 2007). The location of PCMs affect their performance (Al-Absi et al. 2020), and they are generally placed in direct contact with the internal environment of a building (Jin et al. 2016), often in the form of a phase change impregnated plasterboard. In contrast, the analogous biological insulating tissues in animals, such as fur, hair and feathers, are found on the outermost layer of the skin.

PCMs have been viewed as an alternative to thermal mass, as in both strategies heat is stored within the building fabric (see 7.2.2, heat storage).

7.3.3 Controlling heat loss

Increasing heat loss

Ground cooling and heating: See 7.3.2

Evaporative cooling: Evaporative coolers in buildings can be direct and indirect. In both types, water is used for cooling the air, the only difference being in how the air is cooled. In direct cooling, water is used as the cooling medium while in indirect cooling, the outdoor air is cooled by a precooled airflow (Al-Juwayhel et al. 1997). The two parameters of the mass (water) flow rate and packing thickness affect the thermal efficiency of this mechanism. Evaporative cooling from sweat on the skin of some animals is thus like direct evaporative cooling in buildings. Panting also induces evaporative cooling, and thus could be linked to indirect cooling in buildings.

Decreasing heat loss

PCM kinetic thermal insulation: See 7.3.2 for the same strategy.

Avoiding heat loss

Airtightness: The aim of making a building airtight is to control unwanted air movement and consequently, unwanted heat loss. Seals around windows are used to control air leakage. Building wrap is also used to control airflow through a wall or ceiling, although most building wraps are vapour permeable to allow any moisture in the form of water vapour to escape, given the vapour pressure inside a building is generally greater than that outside. There is no direct parallel in the natural world.

7.4 Active and passive methods of thermal regulation in buildings (Plants)

The fact that the movement of plants is limited makes them more like buildings. This section introduces energy-efficient building design strategies that seem equivalent to the mechanisms that plants use for thermal adaptation. As discussed in 6.4, it is not easy to separate active and passive strategies for plants. A large number of the strategies used by plants have already been discussed in 7.2 and 7.3. Given this, cross-references are provided where the strategies have been discussed before.

7.4.1 Generating heat

* *Renewable resources (weak parallel)*: As discussed in 6.3.1, endothermy in plants is not the same as thermoregulation as only limited types of plants sense heat as an external stimulus, although this view has been challenged (Michaletz et al. 2015). Besides, heat generation in both endothermic plants and plants with a thermoregulation capability such as the scared lotus, does not happen through respiration but rather via electron transport at the cellular level. If either endothermic mechanisms or thermoregulation in plants occurred through respiration, this would be a crude parallel with heat generation using renewable energy resources such as PV solar systems and wind turbines.

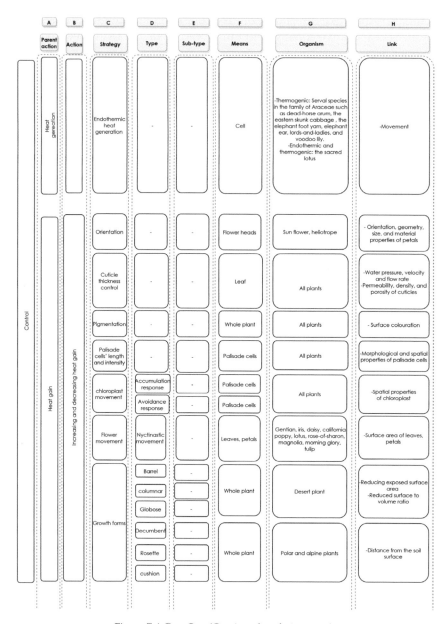

Figure 7-4, Part One (Continued on facing page)

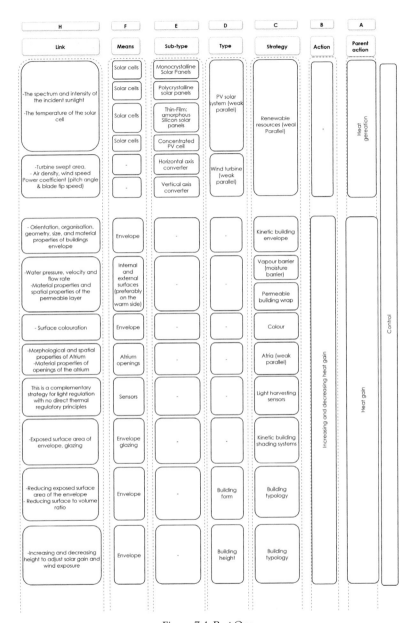

Figure 7-4, Part One

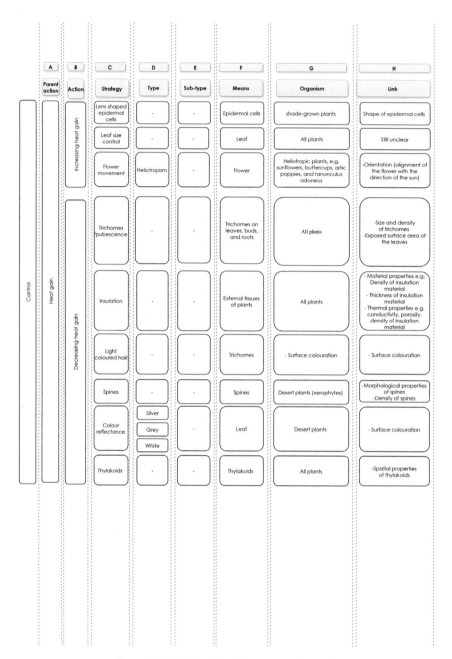

Figure 7-4, Part Two (Continued on facing page)

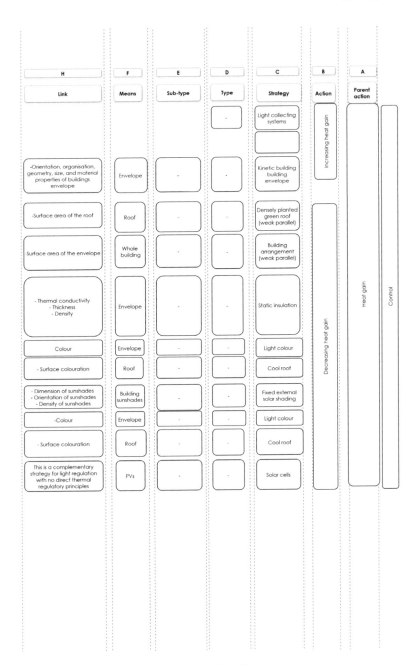

Figure 7-4, Part Two

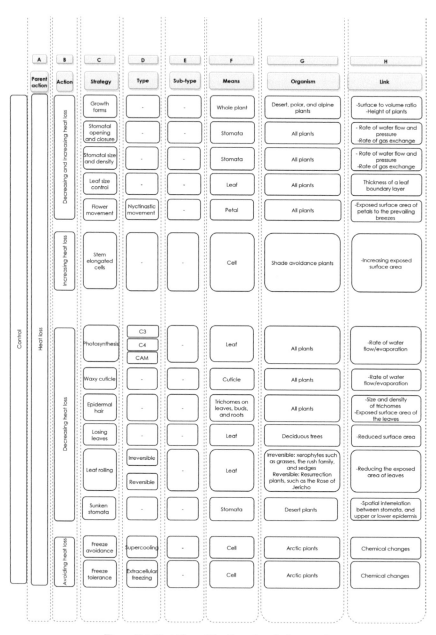

Figure 7-4, Part Three (Continued on facing page)

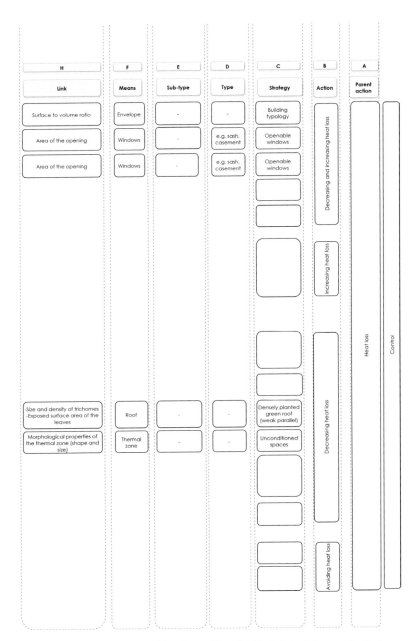

Figure 7-4. Strategies used by plants aligned with their building equivalent.

7.4.2 Controlling heat gain

Increasing and decreasing heat gain

Kinetic building envelope: The kinetic movement of the building envelop seems equivalent to nyctinastic movement in plants. This has been discussed earlier in 7.2.2.

Vapour barrier (moisture barrier) + permeable building wrap: Control of water evaporation from leaves in plants through the presence of the waxy cuticle seems analogous to moisture control in buildings through the use of both vapour barriers and permeable building wrap. A vapour barrier on the inside of the building envelope prevents water movement through it. When water penetrates the envelope, deterioration can occur, which would affect the building life, and also influence the thermal performance of the envelope as damp materials can enhance heat loss by from 2 to about 9 percent (Karagiozis et al. 1995). At the same time, a permeable building wrap on the outside of the envelope will allow any moisture that does enter the envelope, possibly through a failure in the internal vapour barrier, to pass to the outside. In plants, it is not clear whether the thickness of cuticle affects water permeability in leaves (see 6.3.2).

Colour: The role of pigmentation in adjusting solar absorption in plants seems to be similar to that of colour in solar heat gain through the building envelope. Colour affects the absorptivity of the exterior surface of the walls and roof of a building, and the colour of the glass can also affect how much solar energy is transmitted through it (Natephra et al. 2018). While an appropriate colour will not make a building energy-efficient on its own, it is part of the energy balance equation; also refer to the cool roof which has been referred to as a sustainable design method for decreasing heat gain (see 7.2.2).

A type of material that could constantly alter its colour and subsequent levels of solar energy absorption could be conceived as an imaginary parallel to the pigmentation strategy, as plants use different shades of colour to adjust the solar radiation received, although there is contradictory research about whether there is a relationship between floral colour and temperature (see 6.4.2). It has also been suggested that colour could be applied to phase change materials (PCMs) in buildings. While traditional PCMs do not change colour, "the integration of PCMs in coloured containers or with coloured pigments in the PCM could enhance the architectural integration value of PCM technologies" (Vigna et al. 2018). Recent research has demonstrated how the combination of phase-change materials and metamaterial absorber structures could create a novel tuneable colour generation system with potential applications in windows

(Carrillo et al. 2019). The opportunities that such a new generation of PCMs might provide could be analogous to the pigmentation strategy in plants. Developing these could be a step closer to William McDonough's idea of "…why can't I design a building like a tree? A building that changes colours with the seasons and self-replicates" (McDonough 2004).

Atriums (weak parallel): A weak parallel to columnar palisade cells is an atrium. As discussed earlier (see 6.4.2), the size and density of palisade cells regulate light absorption in leaves. Both atria and columnar cells allow light to penetrate deeper into the building interior and the structure of a leaf, respectively. However, there seems to be a difference between these as despite several studies proving the benefits of atriums in reducing artificial lighting demand in a building, the impact of such spaces on the overall thermal load is unclear (Aldawoud 2013). This is in contrast with what happens in a leaf as a result of evolving such cells. As explained in Chapter 6, plants living in an environment with high light intensity develop longer palisade cells in order to reduce light absorption.

Light harvesting sensors: While it seems difficult to foresee the future development of a strategy for designing energy-efficient buildings that replicates chloroplast movements for light regulation, the accumulation and avoidance response of chloroplasts have inspired scientists in other fields. An example would be the creation of design sensors that could read data from chloroplast movements to optimise light conditioning in vertical farms (Kemper 2020). This is because chloroplast movement plays an important role in a plant's energy efficiency (see 6.4.2, palisade cell length and intensity + chloroplast movement). Imitating thylakoids in chloroplast, scientists have created protein nanosheets that structurally resemble thylakoids (Zhao et al. 2017). This bio-inspired technology does not replicate chloroplast movement but instead that of the structure and assemblage of thylakoids (see below in this section for more discussion of solar cells in 'decreasing heat gain').

Kinetic building shading systems: Plant actuators are responsible for moving the organs enabling adaptive responses to external stimuli. Actuation in plants is induced either by external mechanical forces such as wind, the hydraulic processes of swelling and shrinking in cells and tissues and the release of stored elastic energy (Körner et al. 2017). Depending on the types of movement and actuation, a variety of biomimetic structures have been developed.

A parallel to heliotropism could be a kinetic facade that follows the sun path to control heat gain through shading (see 7.2.2, kinetic building envelope), or window blinds that rotate based on the signals they receive from a sensor tracking the sun's position. However, the primary purpose of such systems is normally to improve daylighting. An example is the patented Flectofin blind system, in which blinds are made of fibre-reinforced plastics that imitate the elasticity and deformation mechanisms of plants (Lienhard et al. 2011). In some shading systems, deformation is achieved through the use of smart materials (actuators) that are either separated or integrated into the shading component (Fiorito et al. 2016). A different approach has been the theoretical development of a biomimetic device analogous to a sun-tracking leaf that could be used in photovoltaic panels in place of conventional mechanical tracking systems (Dicker et al. 2014).

Similarly, the structural and anatomical properties plants, including the reversible opening and closing movement of flowers, have inspired speculation about a wide spectrum of possible engineering materials and technical structures (Li and Wang 2016, Poppinga et al. 2018). In a potential application to buildings, the University of Stuttgart *HygroSkin-Meteorosensitive Pavilion* exploits the principle of hygroscopic shape change: "The dimensional instability of wood in relation to moisture content is employed to construct a metereosensitive architectural skin that autonomously opens and closes in response to weather changes" (Krieg 2014). However, this is different from the nyctinastic or day-night opening and closing of flowers. The latter would seem to have less to do with buildings although the traditional use of shutters across windows and the closing of windows at night could be viewed as a weak parallel. In fact, experiments with a biomimetic kinetic envelope have tended to imitate the morphological characteristics of nastic movement (non-directional movement in response to stimuli), or movement due to swelling and shrinkage (Hosseini et al. 2019) as was used in the University of Stuttgart pavilion. It seems that, compared to heliotropism, there is no strong parallel to flower opening and closing in energy-efficient building design. There is some parallel with the prototype folding façade shading system known as Flectofold, which was inspired by the carnivorous waterwheel plant (*Aldrovanda vesiculosa*) (Körner et al. 2017), which snaps shut when its hairs are triggered. In the prototype, movement is activated by use of a pneumatic cushion at the hinge. As the pressure is increased, the leaves fold inwards (Figure 7-5).

Building typology: In cold climates, traditional dwellings were often low in height, such as the single-storey crofts of the Scottish Highlands. This has a parallel with plants in colder climates growing close to the ground to take advantage of the slightly warmer soil temperature and reduce exposure to the wind. Plants will also grow upwards towards the sunshine and here there is a weak parallel with modern high-rise buildings, which try to optimise the land area to the built floor area ratio. In high-rise cities like New York, building facades were stepped back as they rose up to ensure light and fresh air reached street level (Flowers 2009).

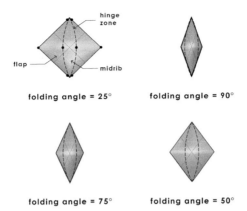

Figure 7-5. Prototype folding façade shading system (Author, adapted from Körner et al. (2017)).

Decreasing heat gain

Densely planted green roof (weak parallel) + building arrangement (weak parallel): Given the primary function of trichomes is absorbing water there is no direct parallel for this in buildings, where the aim is to keep water out of both the building interior and any materials that would be damaged by it. However, as noted in Chapter 6, an increase in the density of trichomes has been linked to a decrease in solar absorption. A weak parallel here would be a densely planted green roof as opposed to one planted with succulents, as the former would offer more potential cooling. Experimental research found that herbaceous plants such as salvia and stachys provided better cooling and insulation than the popular succulents used on a lightweight extensive green roof (Blanusa et al. 2013).

Another weak link with the reduction in solar absorption with an increased density of trichomes comes from the traditionally dense arrangement of buildings in hot climates to reduce the area of walls exposed to the sun (Almaiyah et al. 2010).

Insulation: See static thermal insulation, 7.2.2.

Light colour + cool roof: The obvious building parallel to light-coloured trichomes is the cool roof and the use of light colours to minimise solar absorption, which have already been discussed.

Fixed external solar shading: Spines provide shade so there is an obvious parallel with external shades applied to building facades.

Light colour + cool roof: These are parallel to colour reflectance strategy in plants.

Solar cells: As discussed earlier, the tightly stacked thylakoids containing chlorophylls produce efficient photosynthesis in leaves. To enhance the energy transfer, chlorophylls covering thylakoids need to maintain a fixed distance and orientation. This smart light harvesting mechanism has been explored for its potential in developing strategies for converting and utilising solar energy (Zou et al. 2016). Its possible applications would be use in optical sensors, solar cells, photocatalysis and light emitting materials (Zhao et al. 2017).

Increasing heat gain

Light collecting systems: Given that the main function of lens-shaped epidermal cells is to increase light absorption rather than to regulate heat in a leaf, the technological innovations inspired by such cells are primarily designed to enhance solar collection in order to improve lighting in buildings. Lens-shaped translucent, formations are found on the plant species called *Haworthia Obtusa,* a small succulent. These are known as fenestrate leaves or window leaves as they increase photosynthesis by allow light to reach the centre of the plant, which is rich in chlorophyll.

Replicating the optical characteristics of these window leaves, a novel lighting system has been developed containing a scattering medium that collects and guides solar light to an optical fibre and then carries it through to indoor spaces (Gonome et al. 2020). Also, following a similar approach, new generations of PVs have been proposed for urban environments where illumination levels are low and indirect (Yun et al. 2019). Using a Fresnel lens has also been shown experimentally to enhance the performance of a heat collecting solar array (Lin et al. 2014).

No parallel for leaf size control: As discussed in Chapter 6 the relationship between leaf size and climate is still uncertain. When it comes to traditional buildings, these have tended to be more open in hot, humid climates to allow for air circulation and more closed and compact in colder climates. However, there is no apparent link between the control of the leaf size and increasing heat gain in buildings.

Kinetic building envelope: See 7.2.2, heliotropism.

7.4.3 Controlling heat loss

Increasing and decreasing heat loss

Building typology: Traditional buildings have typically been open to the prevailing breezes to increase heat loss, while to decrease heat loss they are more compact and with fewer openings. The latter have a parallel with plants growing close to the ground in colder climates.

Window operation (weak parallel): A crude parallel to stomatal opening and closure is the opening and closing of windows (see 7.2.2, solar heat gain, kinetic building envelope).

No parallel for leaf stomatal size and density: As discussed in Chapter 6, leaf stomatal density and size can change with a change in temperature. It is hard to see a parallel with buildings. A weak link might be opening windows to increase heat loss on a warm day, and closing them when it is cold. Here it is not the window that is changing in size but the area of opening.

Leaf size control: see 7.4.2, increasing heat gain, no parallel for leaf size control.

No parallel for flower movement: This has no parallels with buildings.

Increasing heat loss

No parallel for stem elongated cells: As noted in Chapter 6, it has been speculated that greater elongation, and hence growth of the stem, is concerned in dissipating heat. This happens when the daytime temperature is higher than that at night. It is hard to see a parallel between stem elongated cells and buildings. Plants also grow upwards in competition for sunlight, although the growth of tall buildings is determined by many other factors than this. There is, perhaps, a weak link to certain building practices in countries like New Zealand, where there are regulations about the placement of buildings to ensure some solar access, although this is more about amenity than increasing heat loss.

Decreasing heat loss

No parallel for photosynthesis: This has no parallels with buildings.

No parallel for waxy cuticle: The waxy cuticle prevents water loss, which in turn would lead to evaporative cooling. Saving water in buildings is an issue but this is approached in other ways such as minimising leaks, installing water efficient appliances and recycling water (Seneviratne 2007).

Green roof (weak parallel): (see 7.4.2. Trichomes).

Unconditioned spaces: Plants lose leaves to reduce surface area, which in turn reduces evaporation and hence, leads to water conservation. The nearest parallel with buildings is reducing heat loss through not heating parts of the building in the cold season and withdrawing to a small heated area, such as the calefactory in a medieval monastery. Although not about increasing heat loss, a similar thing happens in warmer climates where only the occupied room is cooled with a portable air conditioner.

No parallel for leaf rolling: Leaf rolling is a strategy for reducing water loss rather than decreasing heat loss and so has no parallel in buildings.

No parallel sunken stomata: Similar to leaf rolling, the strategy of sunken stomata is for reducing evaporation and hence, water loss, and has no apparent parallel in buildings.

Avoiding heat loss

No parallel for freeze avoidance: Freeze avoidance is a physiological mechanism which happens at the molecular level, and so has no parallel in sustainable building design.

No parallel for freeze tolerance: Like freeze avoidance, there is no obvious parallel with the design of buildings.

7.5 The hierarchical structure of the first draft of the ThBA

Having set out the biological thermal adaptation mechanisms and their architectural equivalents next to each other, it was important to connect them in systematic way. As discussed in Chapter 6, the main categories were named 'parent actions' (column A) consisting of heat gain, heat loss and heat generation, and the sub-branches of these were named 'actions' (column B). In the ThBA, 'actions' branch into 'strategies' (column C), then into 'types' (column D), and finally into sub-types (column E).

The next step was to add more columns to facilitate the bio-inspired design process. This shaped the first draft of the ThBA, also known as ThBA Version 01. For each 'type', there is a 'means' that refers to the part(s) of an animal or plant responsible for thermal adaptation.

Active thermal adaptations in nature happen in a hierarchical order, meaning that physiological and chemical reactions in deeper layers trigger changes in organs. Column F shows the 'means' which are the ultimate organ or tissue involved in active strategies. However, for some strategies and types, only cells deal with thermal adaptation without the rest of the body of the organism being involved. Also, for passive strategies, 'means' are the whole bodies of organisms.

In the ThBA related to animals, an example is introduced for each strategy (column G). Column H is explained below. Figure 7-6 shows the structure of the first draft of the ThBA.

As part of the design by analogy process, to find an analogy in biology, column H named 'link' was generated to provide useful snippets of information for architects to help the translation of thermal adaptation principles found in nature into architectural design principles. The links provide users with a list of the parameters that play a role in the thermal adaptation mechanisms. Such information will assist designers to draw an outline of where and how an aspect of a building could be modified to improve overall building energy efficiency.

Not all principles make an equal contribution to thermal regulation. So, selecting different links could lead to distinct design approaches.

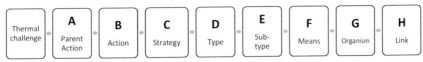

Figure 7-6. Biological information provided in columns A to G in the ThBA framework.

Having said that, it might be feasible to categorise biomimetic building design approaches in the form of a depository for designing energy-efficient buildings. However, to create an inclusive list of these, further research is required which is beyond the scope of this book. Instead, this book aims to explain how a systematic process for finding relevant biological thermal adaptation mechanisms was developed. As part of this, it was necessary to test the ThBA and the next chapter details the steps taken towards developing its final version.

7.6 Biology to architecture transfer

Having looked at parallels between building and the thermoregulation strategies found in animals and plants, there was a need for a mapping stage to facilitate the transfer of knowledge between the domains (Gust et al. 2008). At this stage, the biological solutions should be abstracted to basic principles in order to assist designers in finding appropriate ways for their use in architecture. The following three steps trace the hierarchical structure of creating links between biological mechanisms and architectural design strategies. Looking at thermal adaption as a process in biological organisms, there is always a hierarchical connection descending from

1) The whole body of the organism to
2) the ultimate organ or tissue (the organ or tissue in contact with the environment), and/or the multiple organs/tissues/cells in an interaction, and
3) ultimately between organs, tissues or cells.

These three steps can be abstracted for any active thermal adaptation mechanism such as sweating or panting. Figure 7-5 shows the levels at which the generalisation could occur for both passive and active strategies. More information about how Figure 7-5 was created is provided in Chapter 8.

Although the adaptation mechanism happens in a hierarchical order, some principles seem to play a primary role in homeostasis. For almost all active strategies, the ultimate organ/tissue is central to the adaptation mechanism. However, passive strategies are different as these involve the whole, unchanged body of the organism interacting with its surrounding environment. For the three steps outlined above, the following were the principles on which links were established (Figure 7-7).

4) morphological, spatial (location/position) and material properties for the whole body,

5) morphological, spatial (spatial interrelation of internal structure) and material properties for the ultimate or interrelated organs or tissues or cells affected, and

6) fluid and gas properties relevant to organs/tissues/cells.

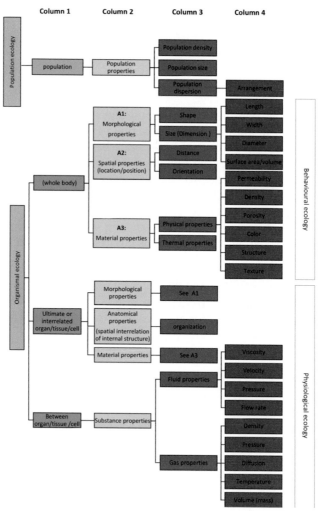

Figure 7-7. The hierarchical structure of mapping links between biological mechanisms and architectural design strategies.

For example, the size and shape of the body of an animal affect thermoregulation in mammals. Body mass and the body surface area both influence the dissipation of heat through the convective heat transfer coefficient. Accordingly, a large mammal hunched up to reduce its surface area would dissipate much less heat than the same mammal not hunched up. Some animals can change their size for other reasons. For example, the tardigrade can lose body water to the extent the skin becomes totally dehydrated (Crowe 1972, Horikawa et al. 2006, Horikawa et al. 2008). In this contracted or suspended animation state, known as the tun state, there is no symptom of life, but the tardigrade can 'come back to life' once rehydrated. This non-metabolic state is termed anhydrobiosis (Keilin 1959); however, it is a survival mechanism rather than a thermoregulation strategy.

7.7 The complementary aspects of thermoregulation

Thus far, thermoregulation mechanisms have been considered in isolation in order to develop the ThBA. However, many thermoregulation mechanisms are used in combination and this section explores this idea further. For example, metabolism, circulation and respiration work together to keep homeostasis in endotherms. For ectotherms that do not generate heat within their bodies, circulation and respiration work together. Complete photosynthesis is the combination of both the physical and chemical aspects of photosynthesis. The fact that systems have to work together to achieve thermoregulation has parallels with the operation of HVAC systems in buildings, and this is explored in this section.

In animals, the cardiovascular system is the way vital oxygen finds its way to the different parts of the body, while also supporting the circulatory system and thermoregulation. The main function of the former is enabling gas exchange between blood and tissues. Blood absorbs oxygen and releases carbon dioxide via the lungs, meaning the circulatory and respiratory systems work together.

7.7.1 Systems in organisms

Respiratory systems: In the respiratory system of an animal like a human being, breathing in happens when muscles in the chest and abdomen contract, causing air to flow into the lungs, whereas when the muscles relax, the reverse happens and the lungs deflate. On the inward breath, the air first goes through the nasal cavity, where it is warmed and also becomes more humid. Then the air passes to the lungs via the throat. At the end of all the bronchial tubes are the alveoli, where gas exchange occurs between the alveoli and the tiny capillary blood vessels that are embedded in their walls. The two factors that affect this gas exchange are the diffusion distance and the surface to volume ratio of the alveoli. Diffusion in cells happens when something moves from a region of high concentration to one of low concentration; a process that can only happen through a thin cell membrane.

In simple, unicellular organisms such as cnidarians and flatworms, and organisms where the flat shape increases the surface area such that all cells in the body are close to the outer membrane, diffusion through this membrane is the means by which respiration occurs (Reece et al. 2015). In larger animals, other arrangements for respiration are necessary; for example, the process in the lungs of mammals is as described above. Additionally, some small organisms such as earthworms and frogs use their skin with its dense network of capillaries for respiration. For this to happen the skin has to remain moist. Creatures that live permanently in water have developed gills. The latter are highly folded to provide a large surface to volume ratio to facilitate gas exchange. All these systems involve blood as a transfer medium.

Respiration in insects does not involve blood. Instead, oxygen enters the body through small openings known as spiracles. These are linked to the trachea in a branching system and it is the pumping of the body segments that moves the gases around.

All respiratory surfaces require a large surface area for gas exchange, a thin membrane for gas diffusion, a moist surface and a concentration gradient such that there is pressure for diffusion to occur.

Table 7-2 sets out the characteristics of three types of respiratory system.

Table 7-2. Respiratory systems in human lungs, fish gills and leaf cells.

System	Large surface area	Small distance	Concentration gradient
Human lungs	600 million alveoli with a total area of 1000 m²	Each alveolus is 1 cell thick	Constant ventilation replaces the air
Fish gills	Feathery filaments with secondary lamellae	Lamellae are 2 cells thick	Water pumped over gills countercurrent to blood
Leaves (tree)	SA of leaves = 200 m² SA of spongy cells inside leaves = 6000 m²	Gases diffuse straight into leaf cells	Wind replaces air around leaves

Circulatory systems: Circulatory systems are both open and closed, while the basic components of each are a network of interconnecting vessels, a pump and a transport fluid.

In an open system, such as is found in grasshoppers and some molluscs, the contraction of the heart creates a pressure causing the fluid (haemolymph) to be sent into the interconnected sinuses or hemocoels, which are spaces surrounding the organs. When the heart relaxes, the fluid is returned to it. The haemolymph is not involved in respiration but is used to carry nutrients and hormones around the body.

In a closed circulatory system, the fluid (blood) is pumped from one or more hearts to first the large and then the small branches of the blood vessels until it penetrates the organs. Annelids (including earthworms), cephalopods (including octopuses and squids) and all vertebrates have a closed circulatory system.

Except for tracheal systems, other systems are closed. In a closed system, circulation and respiration are combined while in an open system, circulation and respiration are separate from each other. In the latter, oxygen is carried by the blood to muscles and tissues, while in the former, the air is taken directly to the tissues of insects. It seems there is no parallel for a tracheal system in buildings. A very weak parallel would be a building in which temperature is controlled through both natural ventilation and zoned HVAC systems that can heat or cool individual areas.

7.7.2 Interconnection of systems

From the preceding description, in a thermoregulatory strategy such as panting, it is obvious that both the respiratory and circulatory systems are necessary for this to happen. The manner in which the transfer of heat in the living human body takes place is complicated, owing to the fact it is a combination of thermal conduction, convection and metabolic heat production. Conduction and diffusion take place in tissues, while convection occurs in blood and other body fluids.

7.7.3 HVAC in buildings and circulatory and respiratory systems in organisms

There are similarities between the circulatory and respiratory systems of animals and the heat transfer mechanisms found in HVAC systems in buildings (Figure 7-8). The simple opening of windows and systems that do not use ducts, such as a split air conditioner with its two parts, one inside and one outside the building, resemble the gas exchange mechanism of unicellular and simple organisms.

Ducted HVAC systems are more like the cardiac system found in more complex organisms. The ducted system can provide heating or cooling by delivering air at the required temperature. A unit containing the condenser, compressor and air handler will be located somewhere convenient, often on the roof or in the basement. Such a centralised system

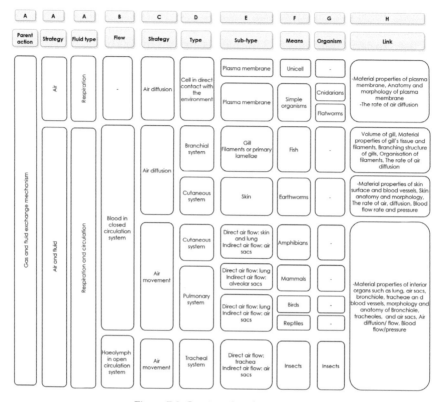

Figure 7-8, Continued on facing page

corresponds to the circulatory systems found in fish, simple organisms such as earthworms, amphibians, birds and mammals.

Heat and cooling can be transferred from the centralised plant to the building using both water and air. This suggests gas exchange in the respiration systems of unicellular and simple organisms can be viewed as analogous to air-cooled HVAC units, whereas the water in water cooled systems plays the same role as that of blood in the circulatory systems of the cutaneous, pulmonary and branchial systems of other organisms. The difference is that in circulatory and respiratory systems in animals, the purpose is to transfer oxygen to the cells, not heat or cooling.

The air and fluid transfer in a circulatory and respiratory system is similar to that of an air- or water-cooled air conditioning system (Figure 7-9 right).

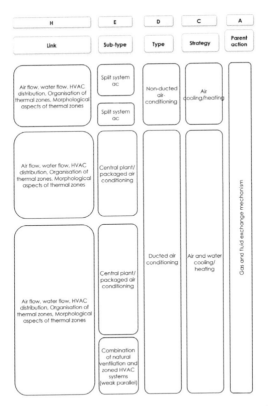

Figure 7-8. (Left) circulatory and respiratory systems in organisms and (Right) HVAC systems in buildings.

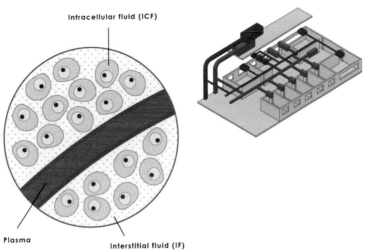

Figure 7-9. (Left) gas exchange mechanism in tissues (Author, adapted from Lumen (2017)); (Right) HVAC systems in a building (Author, adapted from Brown Technical (2018)).

References

Aboulnaga, M. M. (1998). A roof solar chimney assisted by cooling cavity for natural ventilation in buildings in hot arid climates: An energy conservation approach in Al-Ain city. *Renewable Energy, 14*(1-4), 357–363. doi:Doi 10.1016/S0960-1481(98)00090-1.

Al-Absi, Z. A., Mohd Isa, M. H. and Ismail, M. (2020). Phase change materials (PCMs) and their optimum position in building walls. *Sustainability, 12*(4), 1294.

Al-Juwayhel, F. I., Al-Haddad, A. A., Shaban, H. I. and El-Dessouky, H. T. A. (1997). Experimental investigation of the performance of two-stage evaporative coolers. *Heat Transfer Engineering, 18*(2), 21–33. doi:10.1080/01457639708939893.

Aldawoud, A. (2013). The influence of the atrium geometry on the building energy performance. *Energy and Buildings, 57*, 1–5.

Almaiyah, S., Elkadi, H. and Cook, M. (2010, Accessed 10 February 2021). Study on the visual performance of a vascular dwelling in Egypt. *First International Conference on Sustainability and The Future*. Retrieved from http://csfs.bue.edu.eg/files/Library/Papers/Sustainability%20and%20the%20Future/106.pdf.

Alrasheed, M. R. (2011). *A Modified Particle Swarm Optimization and its Application in Thermal Management of an Electronic Cooling System.* (Doctor of philosophy), University of British Columbia, Vancouver, Canada.

Anderson, C., Theraulaz, G. and Deneubourg, J. L. (2002). Self-assemblages in insect societies. *Insectes Sociaux, 49*(2), 99–110. doi:DOI 10.1007/s00040-002-8286-y.

Awad, M. and Khanna, R. (2015). Bioinspired computing: Swarm intelligence. pp. 105–125. *In*: M. Awad and R. Khanna (eds.). *Efficient Learning Machines—Theories, Concepts, and Applications for Engineers and System Designers.* Berkeley, CA: Apress.

Baetens, R., Jelle, B. P., Thue, J. V., Tenpierik, M. J., Grynning, S., Uvsløkk, S. and Gustavsen, A. (2010). Vacuum insulation panels for building applications: A review and beyond. *Energy and Buildings, 42*(2), 147–172.

Balouria, R. (2019). Honeybees. Retrieved from https://pixabay.com/photos/honeybees-beehive-beekeeping-4060349/, Pixabay License: https://pixabay.com/service/license/.

Beigli, F. and Lenci, R. (2016). Underground and semi underground passive cooling strategies in hot climate of Iran. *Journal of Environmental Science, 5*(3), 198–209.

Bellia, L., De Falco, F. and Minichiello, F. (2013). Effects of solar shading devices on energy requirements of standalone office buildings for Italian climates. *Applied Thermal Engineering, 54*(1), 190–201.

Blanusa, T., Monteiro, M. M. V., Fantozzi, F., Vysini, E., Li, Y. and Cameron, R. W. (2013). Alternatives to Sedum on green roofs: can broad leaf perennial plants offer better 'cooling service'? *Building and Environment, 59*, 99–106.

Brown Technical. (2018). How an HVAC System Works. Retrieved 8 October 2018, from Brown Technical, https://www.browntechnical.org/pages/learning-center-for-hvac/how-an-hvac-system-works.html.

Carrillo, S. G-C., Trimby, L., Au, Y-Y., Nagareddy, V. K., Rodriguez-Hernandez, G., Hosseini, P., Carlos, R., Bhaskaran, H. and Wright, C. D. (2019). A nonvolatile phase-change metamaterial color display. *Advanced Optical Materials, 7*(18), 1801782.

Crowe, J. H. (1972). Evaporative water loss by tardigrades under controlled relative humidities. *Biological Bulletin, 142*(3), 407–416.

Dicker, M., Rossiter, J., Bond, I. and Weaver, P. (2014). Biomimetic photo-actuation: sensing, control and actuation in sun-tracking plants. *Bioinspiration & Biomimetics, 9*(3), 036015.

Efficient Windows Collaborative. (2021). Windows for high-performance commercial buildings. Retrieved from https://www.commercialwindows.org/adv_glass.php.

El Razaz, Z. (2010). Sustainable vision of kinetic architecture. *Journal of Building Appraisal, 5*(4), 341–356.

Fiorito, F., Sauchelli, M., Arroyo, D., Pesenti, M., Imperadori, M., Masera, G. and Ranzi, G. (2016). Shape morphing solar shadings: A review. *Renewable and Sustainable Energy Reviews, 55*, 863–884.

Florides, G. and Kalogirou, S. (2007). Ground heat exchangers—A review of systems, models and applications. *Renewable Energy, 32*(15), 2461–2478.

Flowers, B. (2009). *Skyscraper: The Politics and Power of Building New York City in the Twentieth Century.* University of Pennsylvania Press.

Fox, M. (2016). *Interactive Architecture: Adaptive World.* New York, NY: Princeton Architectural Press.

Givoni, B. (2007). Cooled soil as a cooling source for buildings. *Solar Energy, 81*(3), 316–328. doi:10.1016/j.solener.2006.07.004.

Gonome, H., Watanabe, K., Nakamura, K., Kono, T. and Yamada, J. (2020). Lighting system bioinspired by Haworthia obtusa. *Scientific Reports, 10*(1), 1–7.

González-Espada, W. J., Bryan, L. A. and Kang, N.-H. (2001). The intriguing physics inside an igloo. *Physics Education, 36*(4), 290.

Gusmão, L. C., Van Deusen, V., Daly, M. and Rodríguez, E. (2020). Origin and evolution of the symbiosis between sea anemones (Cnidaria, Anthozoa, Actiniaria) and hermit crabs, with additional notes on anemone-gastropod associations. *Molecular Phylogenetics and Evolution, 148*, 106805.

Gust, H., Krumnack, U., Kühnberger, K.-U. and Schwering, A. (2008). Analogical reasoning: A core of cognition. *KI–Zeitschrift Künstliche Intelligenz, 22*(1), 8–12.

Halford, C. K. and Boehm, R. F. (2007). Modeling of phase change material peak load shifting. *Energy and Buildings, 39*(3), 298–305.

Horikawa, D. D., Sakashita, T., Katagiri, C., Watanabe, M., Kikawada, T., Nakahara, Y., Hamada, N., Wada, S., Funayama, T., Higashi, S., Kobayashi, Y., Okuda, T. and Kuwabara, M. (2006). Radiation tolerance in the tardigrade Milnesium tardigradum. *International Journal of Radiation Biology, 82*(12), 843–848.

Horikawa, D. D., Kunieda, T., Abe, W., Watanabe, M., Nakahara, Y., Yukuhiro, F., Sakashita, T., Hamada, N., Wada, S., Funayama, T., Katagiri, C., Kobayashi, Y., Higashi, S. and Okuda, T. (2008). Establishment of a rearing system of the extremotolerant tardigrade Ramazzottius varieornatus: a new model animal for astrobiology. *International Journal of Radiation Biology, 8*(3), 549–556.

Hosseini, S. M., Mohammadi, M., Rosemann, A., Schröder, T. and Lichtenberg, J. (2019). A morphological approach for kinetic façade design process to improve visual and thermal comfort. *Building and Environment, 153*, 186–204.

Jin, X., Medina, M. A. and Zhang, X. (2016). Numerical analysis for the optimal location of a thin PCM layer in frame walls. *Applied Thermal Engineering, 103*, 1057–1063.

Karagiozis, A., Salonvaara, M. and Kumaran, M. (1995). The effect of waterproof coating on hygrothermal performance of high-rise wall structure. *Thermal Performance of the Exterior Envelopes of Buildings VI, Clearwater, FL-USA.*

Keilin, D. (1959). The problem of anabiosis or latent life: History and current concept. *Proceedings of the Royal Society of B: Biological Sciences, 150*(939), 149–191.

Kemper, P. (2020). *Engineering a Chloroplast Movement Sensor.* University of Twente.

Khan, N., Su, Y. and Riffat, S. B. (2008). A review on wind driven ventilation techniques. *Energy and Buildings, 40*(8), 1586–1604.

Khudhair, A. M. and Farid, M. M. (2004). A review on energy conservation in building applications with thermal storage by latent heat using phase change materials. *Energy Conversion and Management, 45*(2), 263–275. doi:10.1016/S0196-8904(03)00131-6.

Kocer, C. (2020). The Past, Present, and Future of the Vacuum Insulated Glazing Technology. Retrieved from https://www.glassonweb.com/article/past-present-and-future-vacuum-insulated-glazing-technology.

Körner, A., Born, L., Mader, A., Sachse, R., Saffarian, S., Westermeier, A. S., Poppinga, S., Bischoff, M., Gresser, G. T., Milwich, M., Speck, T. and Knippers, J. (2017). Flectofold—a biomimetic compliant shading device for complex free form facades. *Smart Materials and Structures, 27*(1), 017001.

Krieg, O. D. (2014). HygroSkin-Meteorosensitive Pavilion. pp. 125–137. *In:* Achim Menges, Tobias Schwinn and Krieg, O. D. (eds.). *Advancing Wood Architecture: A Computational Approach.* Abingdon, Oxon: Routledge.

Lee, S. W., Lim, C. H. and Salleh, E. (2016). Reflective thermal insulation systems in building: A review on radiant barrier and reflective insulation. *Renewable and Sustainable Energy Reviews, 65,* 643–661.

Li, S. and Wang, K. (2016). Plant-inspired adaptive structures and materials for morphing and actuation: a review. *Bioinspiration & Biomimetics, 12*(1), 011001.

Lienhard, J., Schleicher, S., Poppinga, S., Masselter, T., Milwich, M., Speck, T. and Knippers, J. (2011). Flectofin: a hingeless flapping mechanism inspired by nature. *Bioinspiration & Biomimetics, 6*(4), 045001. doi:10.1088/1748-3182/6/4/045001.

Lin, M., Sumathy, K., Dai, Y. and Zhao, X. (2014). Performance investigation on a linear Fresnel lens solar collector using cavity receiver. *Solar Energy, 107,* 50–62.

Lumen. (2017). Retrieved 26 March 2019, from Lumen Learning https://courses. lumenlearning.com/suny-ap2/chapter/body-fluids-and-fluid-compartments-no-content/.

McDonough, W. (2004) *Thinking like Nature: William Mcdonough Redesigns the World/ Interviewer: D. Pollard.*

Michaletz, S. T., Weiser, M. D., Zhou, J., Kaspari, M., Helliker, B. R. and Enquist, B. J. (2015). Plant thermoregulation: energetics, trait–environment interactions, and carbon economics. *Trends in Ecology & Evolution, 30*(12), 714–724.

Morakinyo, T. E., Dahanayake, K. K. C., Ng, E. and Chow, C. L. (2017). Temperature and cooling demand reduction by green-roof types in different climates and urban densities: A co-simulation parametric study. *Energy and Buildings, 145,* 226–237.

Natephra, W., Yabuki, N. and Fukuda, T. (2018). Optimizing the evaluation of building envelope design for thermal performance using a BIM-based overall thermal transfer value calculation. *Building and Environment, 136,* 128–145.

Niglio, M. (2018). *Past as Future in Adaptive Buildings: Climatic Adaptation in Ancient Constructions.* Paper presented at the Advanced Materials Research.

Nikolaou, T., Kolokotsa, D., Stavrakakis, G., Apostolou, A. and Munteanu, C. (2015). *Managing Indoor Environments and Energy in Buildings with Integrated Intelligent Systems.* Cham, Switzerland: Springer.

Oberndorfer, E., Lundholm, J., Bass, B., Coffman, R. R., Doshi, H., Dunnett, N., Gaffin, S., Köhler, M., Lui, K. K. Y. and Rowe, B. (2007). Green roofs as urban ecosystems: ecological structures, functions, and services. *BioScience, 57*(10), 823–833.

Ocko, S. A. and Mahadevan, L. (2014). Collective thermoregulation in bee clusters. *Journal of the Royal Society Interface, 11*(91), 1–14. doi:10.1098/rsif.2013.1033.

Parameshwaran, R., Kalaiselvam, S., Harikrishnan, S. and Elayaperumal, A. (2012). Sustainable thermal energy storage technologies for buildings: A review. *Renewable & Sustainable Energy Reviews, 16*(5), 2394–2433. doi:10.1016/j.rser.2012.01.058.

Pino, A., Bustamante, W., Escobar, R. and Pino, F. E. (2012). Thermal and lighting behavior of office buildings in Santiago of Chile. *Energy and Buildings, 47,* 441–449. doi:10.1016/j. enbuild.2011.12.016.

Poppinga, S., Zollfrank, C., Prucker, O., Rühe, J., Menges, A., Cheng, T. and Speck, T. (2018). Toward a new generation of smart biomimetic actuators for architecture. *Advanced Materials, 30*(19), 1703653.

Ramzy, N. and Fayed, H. (2011). Kinetic systems in architecture: New approach for environmental control systems and context-sensitive buildings. *Sustainable Cities and Society, 1*(3), 170–177. doi:10.1016/j.scs.2011.07.004.

Ratti, C., Baker, N. and Steemers, K. (2005). Energy consumption and urban texture. *Energy and Buildings, 37*(7), 762–776. doi:DOI 10.1016/j.enbuild.2004.10.010.

Reece, J. B., Meyers, N., Urry, L. A., Cain, M. L., Wasserman, S. A., Minorsky, P. V., Jackson, R. B., Cooke, B. and Campbell, N. A. (2015). *Campbell Biology: Australia and New Zealand Edition* (10th ed.). Melbourne, Victoria: Pearson Australia.

Roberts, S. (2008). Effects of climate change on the built environment. *Energy Policy, 36*(12), 4552–4557. doi:10.1016/j.enpol.2008.09.012.

Rosado, P. J., Faulkner, D., Sullivan, D. P. and Levinson, R. (2014). Measured temperature reductions and energy savings from a cool tile roof on a central California home. *Energy and Buildings, 80,* 57–71.

Seneviratne, M. (2007). *A Practical Approach to Water Conservation for Commercial and Industrial Facilities.* Elsevier.

Smith, C. S. (2017, 22 June). A lost art in the Arctic: Building an Igloo. *The New York Times,* A7.

Torgal, F. P. (2016). Introduction to nano-and biotech-based materials for energy building efficiency. pp. 1–16. *In*: T. F. Pacheco, C. Buratti, S. Kalaiselvam, C. G. Granqvist and A. G. Ivanov (eds.). Nano and Biotech Based Materials for Energy Building Efficiency. Springer.

Tzempelikos, A. and Athienitis, A. K. (2007). The impact of shading design and control on building cooling and lighting demand. *Solar Energy, 81*(3), 369–382. doi:10.1016/j.solener.2006.06.015.

Vigna, I., Bianco, L., Goia, F. and Serra, V. (2018). Phase change materials in transparent building envelopes: A Strengths, Weakness, Opportunities and Threats (SWOT) analysis. *Energies, 11*(1), 111.

Whitridge, P. (2008). Reimagining the Iglu: Modernity and the challenge of the eighteenth century Labrador Inuit winter house. *Archaeologies, 4*(2), 288–309.

Yang, X., Chen, Z., Cai, H. and Ma, L. (2014). A framework for assessment of the influence of China's urban underground space developments on the urban microclimate. *Sustainability, 6*(12), 8536–8566.

Yun, M. J., Sim, Y. H., Cha, S. I. and Lee, D. Y. (2019). Leaf anatomy and 3-D structure mimic to solar cells with light trapping and 3-D arrayed submodule for enhanced electricity production. *Scientific Reports, 9*(1), 1–9.

Zhao, L., Zou, H., Zhang, H., Sun, H., Wang, T., Pan, T., Li, X., Bai, Y., Qiao, S., Luo, Q., Xu, J., Hou, C. and Liu, J. (2017). Enzyme-triggered defined protein nanoarrays: Efficient light-harvesting systems to mimic chloroplasts. *ACS Nano, 11*(1), 938–945.

Zou, Q., Liu, K., Abbas, M. and Yan, X. (2016). Peptide-modulated self-assembly of chromophores toward biomimetic light-harvesting nanoarchitectonics. *Advanced Materials, 28*(6), 1031–1043.

Chapter 8
Testing the ThBA

8.1 Introduction

Having categorised the thermal adaptation mechanisms of organisms and their equivalent strategies in energy-efficient building design, this chapter tests the first draft of the ThBA. The aim is to see whether it is set out in such a way that it allows architects and designers to finding relevant inspiration in nature, either based on the thermal challenges identified for an existing building or to make energy-efficient decisions at the early design stage.

However, it was necessary that the ThBA was evaluated by biologists. This was done through conducting a focus group, a brief description of which is provided in 8.2. The main purpose of this step was to have the quality, inclusiveness and applicability of the biological theory of the ThBA assessed by experts in the field.

In addition, since the intention was the ThBA would be a roadmap for connecting thermal challenges in a building to relevant biological thermoregulatory solutions in a systematic way, it was important to develop a process through which the thermal behaviour of a building could be understood. What a thermal challenge means in the context of this book is locating where a high energy use occurs in a particular design, since this shows the need for considerable cooling energy or heating energy. The thermal issues in this sense form the input to the ThBA for finding relevant thermal adaptation solutions in nature.

To articulate how thermal challenges in a building can be identified, energy simulations were run for five cases studies to find the two buildings with the most thermal issues. Also, an appropriate approach was developed to narrow down the energy results in a way such that the main thermal challenges along with the heat balance characteristics of a building could be recognised. A brief explanation of the process is provided in Section 8.3.

The energy models used for this book were created based on real building energy use data. Before being used for testing the ThBA, the energy performances of the models were evaluated for matching with those of the real buildings.

8.2 Focus group

To be able to evaluate the inclusiveness of the ThBA, it was necessary to have the opinions of biologists with a range of expertise. There appeared to be eight different biological fields which might generate different approaches to assessing the ThBA. These were zoology, including entomology (study of insects), herpetology (study of reptiles and amphibians), ichthyology (study of fish), malacology (study of molluscs), mammalogy (study of mammals), ornithology (study of birds) as well as botany and microbiology.

The eventual focus group covered a diverse range of expertise including botany, and marine and terrestrial animal biology. This meant the entire system of animal classification was represented as there was an approximate equal distribution of expertise between those working on vertebrates and invertebrates at different biological scales from cells to tissues, organs and single organisms.

The results showed the participants had a good understanding of the ThBA and its application to building design. The content and structure of the ThBA were approved, and the group confirmed no obvious strategies were missing from the list. Also, the themes that emerged from reviewing the thermal physiology of animals that were used to develop the ThBA were corroborated as valid and relevant. Detailed information about the findings of the focus group can be found elsewhere (Imani 2020, Imani and Vale 2020a).

The biologists noted that thermal adaptation strategies take place at different scales and that there might not be links between them for all strategies. They added that the thermal adaptation strategies that happen independently at different scales have the potential to offer different solutions and thus need to be studied carefully.

Another strategy that emerged in Chapter 5 was the close relationship between thermal stressors. In support of this, the focus group referred to complementary environmental stressors, where responding to one would result in thermal adaptation. The stressors mentioned were light, water, nutrients and predators. Also, the participants emphasised that not all biological thermal adaptation strategies are energy efficient. They also agreed there is an energy cost in thermal adaptation strategies and organisms make compromises regarding this all the time.

Overall, the group felt that buildings seemed to be more like plants than animals. For the strategies in building design that had no parallels in biology, the biologists suggested searching for non-documented species, something that is mainly done for producing pharmaceutical products.

Having confirmed the inclusiveness and the structural hierarchy of the ThBA, the next step was to test its effectiveness in suggesting innovative solutions for designing energy-efficient buildings.

8.3 Identification of thermal issues

The case studies used to test the ThBA were in Auckland and Dunedin, New Zealand. The selection of these locations covered the climate extremes of this country. Details about the case study selection and the process used for examining their reliability are discussed elsewhere (Imani and Vale 2020b).

The initial case studies were five office buildings from which two were selected for further energy simulation and thermal performance analysis (Imani 2020), known as buildings A and B (Figure 8-1). These two office buildings created four case studies: Building A in Auckland, Building A in Dunedin, Building B in Auckland and Building B in Dunedin. Analysing the thermal performance for these was carried out at the zone scale to identify the location and duration of the thermal challenges (Table 8-1).

For each case study, the heat balance characteristics of the building were identified from the relevant calculations as non-numeric representations (Table 8-2). Recognition of the reasons behind high energy use in a specific zone and at a specific time of year could assist in identification of the most effective way to improve the energy performance of a building.

The thermal issues and characteristics determined the relevant 'actions' that were then used as input to the ThBA Version 01. 'Actions' were thus the starting point to search the vast range of biological thermal adaptation strategies used by plants and animals. As this book uses the energy models of real office buildings, only the biological solutions whose translation could be applied to an existing building were considered. While using the ThBA Version 01 for designing an energy-efficient building from scratch could suggest some innovative design ideas, this is an area that remains to be explored. Section 8.7 expands the idea of the solution finding process and how solutions might be applied to change the design of a building.

Building A Site: Dunedin, New Zealand			Building B Site: Auckland, New Zealand		
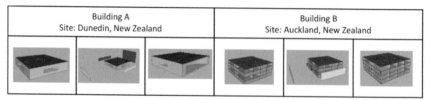

Figure 8-1. Visual thumbnails of buildings A and B.

Table 8-1. Location and duration of thermal challenges (G = ground, M = intermediate and T = top floor).

	A			B								
	Auckland		Dunedin	Auckland			Dunedin					
Thermal challenge	Cooling	Heating	Heating	Cooling			Cooling			Heating		
Season	Summer	Winter	Winter	Summer			Summer			Winter		
Floor	G	G	G	G	M	T	G	M	T	G	M	T
Zones	NE, NW	Peripheral Zones	Peripheral zones	All	All	All	All	All	All	SW, NW, NE, SE	SW, NW, NE, SE	All
Most challenging zones	NE					SW, NW, NE, SE	NE, SE, NEM, NWM	NE, SE, NEM, NWM	NE, SE, NEM, NWM	SE	SE	
Month	Jan, Feb, Mar, Oct, Nov, Dec		Apr, May, Jun, Jul, Aug	Jan, Feb, Mar, Nov, Dec	Jan, Feb, Mar, Nov, Dec	Jan, Feb, Mar, Nov, Dec	Jan, Feb, Mar, Oct, Nov, Dec	Jan, Feb, Mar, Oct, Nov, Dec	Jan, Feb, Mar, Oct, Nov, Dec	Jun, Jul, Aug	Jun, Jul, Aug	Jun, Jul, Aug

Table 8-2. Heat balance characteristics.

A	A	B	B
Auckland	**Dunedin**	**Auckland**	**Dunedin**
External heat gain	External heat gain	Internal heat gain and external heat gain	Internal heat gain and external heat gain
Conductive heat loss through opaque surfaces	Conductive heat loss through opaque surfaces	Conductive heat loss through windows	Conductive heat loss through opaque surfaces and windows
Infiltration heat loss through windows	Infiltration heat loss through windows		
Conductive and radiative heat gain from windows	Conductive and radiative heat gain from windows	Conductive and radiative heat gain from windows	Conductive and radiative heat gain from windows

From the four options generated, the simplest case study was used for testing the ThBA Version 01. This meant the case study building with the fewest thermal challenges to occur over a comparatively short period. Building A in Dunedin had only one thermal challenge (heating) and was selected as the most straightforward case study (Table 8-1). Even though Building A in Auckland also had one thermal challenge (cooling), it was a larger building, and therefore more complex. Building A in Auckland had more challenging thermal zones, as for some zones the cooling energy was considerably higher than for others. Being a four-storey building with high energy use in January, February, March, April and May made it a second priority for testing the ThBA.

To assess its robustness, the plan was to test the ThBA twice, first for building A in Dunedin (Test 01), and then for the same building in Auckland (Test 02). For these two case studies, the inputs were:

1) the thermal issues identified (Table 8-1), and
2) their heat balance characteristics (Table 8-2).

Testing in a second climate was important in order to investigate whether the same building would still perform in different parts of New Zealand. If the results were negative, the intention was to test the ThBA for a second time for finding relevant biological thermal adaptation mechanisms that would produce a building that did perform well in the different New Zealand location.

8.4 Building A in Dunedin (using the ThBA Version 01, Test 01): the need to redesign the ThBA

Looking at heat balance characteristics of buildings in Table 8-2, it appears Building A in Dunedin was a skin-load dominated building as conductive and radiative heat gains through the windows were more significant than internal heat gains. There were also heat losses through the opaque surfaces and the windows with the former being nearly three times that of the latter.

As shown in Table 8-2, for Building A in Dunedin, the thermal challenge was heating. Heating was thus used as the input to the ThBA. The relevant actions for overcoming heating were: 'increasing heat gain' and 'decreasing heat loss'; 'preventing heat loss' and 'stealing heat' seemed similar to 'decreasing heat loss' and 'increasing heat gain'. Likewise, 'generating heat' was considered to be relevant. Using these relevant actions, the next step was to go to the ThBA and look for the thermal adaptation mechanisms each action suggested.

When selecting 'increasing heat gain' as the first action and subsequent search for relevant biological solutions, it seemed the ThBA was not very well organised for assisting the search for solutions. While the biological side of the ThBA version 01 was considered well organised by the biologists, it was soon apparent that it was of no use for architectural applications. This necessitated a reorganisation of the ThBA before looking further for relevant actions, and this process is explained in the following sections.

8.4.1 The process of redesigning the ThBA Version 01

Using the ThBA Version 01, the first step was to select strategies related to 'increasing heat gain' from the related passive strategies for animals and plants. The same process needed to be repeated for active strategies for animals and plants. However, an issue arose when looking at passive strategies for animals. Figure 8-2 shows the relevant strategies that were selected and marked up in the ThBA with a black boundary. These were then imported into a table (Table 8-3). Apart from the inputs of columns A–H (the labelling system of the ThBA Version 01, see Figure 8-2), two more columns were added at the start of the table to show whether the data was related to animals (A) or plants (P), and if the mechanism was passive (PA) or active (AC). Note: This table only deals with animals and passive mechanisms.

The following points emerged from Table 8-3:

1) The 'parent action' was such a broad category that is was not necessary and was removed.

2) The hierarchical structure was reordered to make it more useful for architects. The two columns labelled 'Strategy' and 'Type' would be more useful if combined into a single column labelled 'Solutions'. This was because many strategies had only one type. Doing this would condense the information and facilitate interpretation of the data.

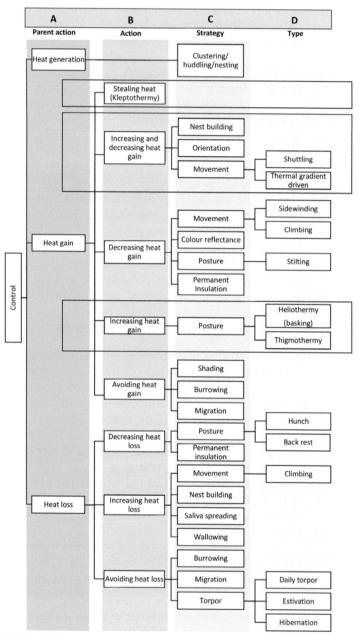

Figure 8-2. The ThBA Version 01 related to passive strategies in animals (used as an example).

Table 8-3. The ThBA Version 01 data for passive strategies of 'increasing heat gain' for animals.

A/P	PA/AC	A Parent action	B Action	C Strategy	D Type	E Sub-type	F Means	G Organism	H link
A	PA	Heat gain	Increasing heat gain	Nest building	-	-	Construction	Bees, termites, ants	- Population size - Material properties of nests - Orientation, geometry of nest and nest openings
A	PA	Heat gain	Increasing heat gain	Orientation	-	-	Whole body	Soma animals	Body axis alignment or misalignment with the direction of the sun
A	PA	Heat gain	Increasing heat gain	Movement	shuttling	-	Whole body	Several lizards	-Maximising both the duration and the surface area of contact
A	PA	Heat gain	Increasing heat gain	Movement	Thermal gradient driven	-	Whole body	Aquatic and semi-aquatic amphibians	Increasing/decreasing distance from the heat source
A	PA	Heat gain	Increasing heat gain	Posture	heliothermy/ basking	-	Whole body	Snakes	- Intelligent body adjustment
A	PA	Heat gain	Increasing heat gain	Posture	thigmothermy	-	Whole body	Several lizards	- Material properties and temperature difference between the animal's body and a hot surface

3) The 'Means' column showed whether an organism undertook a particular thermal adaption through its whole body, organs, tissues or cells. However, this did not seem important for architects and was removed.

4) The 'Organism' column also seemed unnecessary for architects at this stage, as seeking design options seemed more important than examples of animals or plants. So, column 'G' was removed at this stage. In future, it might be useful to prepare a version of the ThBA with this information to enable a more thorough investigation of biological solutions and organisms by those interested in doing this.

5) For each solution, the 'Link' column (column 'H' in Table 8-3) lists the parameters affecting the thermal regulation mechanism. As discussed in Chapter 6, Figure 8-3 was created to assist in mapping the links into a hierarchical structure through which they could become generalised. Even though the adaptation mechanism in a natural organism occurs in a hierarchical order, some principles seem to play a primary role in homeostasis. Therefore, it would be necessary to list the parameters central to each thermal adaptation mechanism. An example of this for vasoconstriction is shown in Table 8-4.

To identify the most useful links, the parameters were reviewed to determine where they fitted within the general categories. The example below shows what happens when this was done for the 'movement' strategy. This was then further tested for being generalised to broader categories, following the hierarchical connection of thermal regulation shown in Figure 8-3.

For 'movement' there were two 'types' from which 'shuttling' was selected and reviewed for its links with regard to:

a) maximising both the duration and,

b) the surface area of contact.

From the two parameters above, only 'b)' seemed appropriate for an architect and thus, 'surface area' was sublimated to 'size' and then to 'morphological properties'.

In 'shuttling', the heat is transferred from the air to the skin of animal. Given that conduction, convection and radiation are the three heat transfer mechanisms governing both living things and building operation, it seemed to be important to document the method of heat transfer that enabled homeostasis while also looking for the key parameters in the heat transfer equation.

Convection is described by:

$$q = hA * (T_1 - T_2) \qquad W$$

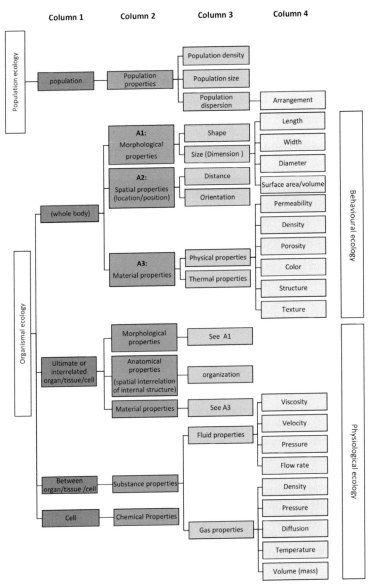

Figure 8-3. The hierarchical connection of thermal regulatory mechanisms in biology.

$$q'' = h * (T_1 - T_2) \qquad W/m^2$$

where h is the convective heat transfer coefficient in Wm^2K, and A the surface area where convection happens in m^2. For 'shuttling', 'surface area' was a key principle for both the biological thermoregulatory mechanism and the convective heat transfer equation.

While the 'shuttling' strategy had few thermoregulatory principles, some solutions in the ThBA seemed to have several of these, out of which only some were central to thermal adaptation. To investigate how links could be developed for complicated strategies, the ThBA sections related to animals and plants were searched again and 'vasoconstriction' was selected as an example of such a solution. Using 'vasoconstriction' as an active solution for decreasing heat loss in animals shows how the main thermoregulatory principles can be determined for complicated mechanisms with hierarchical connections. The links for 'vasoconstriction' (Table 8-4) are:

a) Diameter and length of blood vessels.

b) Permeability, porosity, density and thermal conductivity of the superficial body tissue (skin) and blood vessels.

c) Pressure and blood flow.

The hierarchical structure developed in Chapter 8 (Figure 8-3), was used to generalise the thermoregulation principles into broader categories. There are the two broad categories of physical and chemical properties from which the former branches into mechanical, thermal, electrical, magnetic, and optical, while the latter consists of the two categories of environmental and chemical stability (Aran 2007).

Given this, item 'a)' was generalised to size, item 'b)' to physical properties and item 'c)' to fluid properties. These were linked for a second time, with 'size' being related to 'morphological properties', 'physical properties' to 'material properties' and 'fluid properties' to 'substance properties'. However, the main principle behind vasoconstriction was the change in the diameter of blood vessels, and the consequent change in blood flow. This meant, the main vasoconstriction parameters were 'a)' and 'c)'.

For 'a)' and 'c)', increasing the diameter of a blood vessel seemed to be a primary principle as decreasing the blood flow rate depends on the diameter of blood vessels (Column 4, Table 8-4). For the former, size, and for the latter, fluid properties, change during thermoregulation (Column 3). Size fitted into the broader category of 'morphological properties' and 'fluid properties' was grouped under 'gas exchange

Table 8-4. The links for vasoconstriction.

Level of thermoregulation	Links		
Column 1	Column 2	Column 3	Column 4
Ultimate or interrelated tissues	Morphological properties	Size	Diameter of blood vessel
Between tissues	Gas exchange substance properties	Fluid properties	Blood flow rate

substance properties'. This shows how the links for 'vasoconstriction' were generalised into a broader category at three levels (Table 8-4). This means that moving from Column 4 towards Column 2, the categories gradually become broader. Column 1 shows the level at which the thermoregulatory mechanism takes place. For 'shuttling', thermoregulation happened at ultimate/interrelated tissues and also between them (Column 1, Table 8-4). Matching grey shading is used to show the positioning of the principles related to the 'shuttling' strategy (Table 8-4) in the hierarchical structure (Figure 8-3).

In addition to generalising the main thermoregulatory principles, it was important to determine the type(s) of heat transfer methods and also the main parameter(s) changing in the relevant equation. The information listed in Column 1 assisted in determining these.

Convection involves the transfer of heat through the flow of a fluid over a solid boundary. For 'vasoconstriction', thermoregulation happens through convective heat transfer. There are two types of 'natural' and 'forced' convection. For the former, convection happens through a still fluid while in the latter, the fluid flow is controlled by a pump. Given this, vasoconstriction belongs to the latter category, as the heart pumps the blood into the blood vessels.

For both natural or forced convection, the heat transfer equation has the parameter of 'heat transfer coefficient', the value of which varies based on the fluid flow conditions such as fluid velocity, viscosity and heat flux. For vasoconstriction, the 'heat transfer coefficient' changes with the rate of blood flow to allow the heat transfer between the blood and the tissue, and so the main parameter in the equation is the 'heat transfer coefficient' (Table 8-5).

Therefore, the two parameters of 'heat transfer methods' and 'the main parameter in the heat transfer equation' were identified as important and were thus considered for use in reorganising the ThBA (Table 8-7).

Table 8-5. The links and heat transfer method and variables for vasoconstriction.

Solution	Links		Heat transfer method	Variable in the equation
	Column 4	Column 2		
What is the solution?	**What principle(s) is (are) central to thermal adaptation?**	**What properties of a living organism do the central thermal adaptation principles belong to?**	**What method of heat transfer allows thermoregulation?**	**Which parameter in the heat transfer equation is changed?**
Vasoconstriction	Diameter of blood vessel	Morphological properties	Convective heat transfer	Convective heat transfer coefficient
	Blood flow rate	Gas exchange substance properties		

The duration for which a biological solution was used by the organism seemed to be another useful piece of information for architects. This was the last and sixth point to emerge from Table 8-3 (points 1–5 are found immediately under Table 8-3).

6) As heating energy for Building A in Dunedin was a temporary thermal challenge in winter, it was also important to document whether the solutions were employed temporarily or permanently by organisms.

Given the six points outlined above, including points 1–5 under Table 8-3, the ThBA was reorganised to facilitate identification of relevant thermal adaptation solutions for architects (Table 8-7). Presenting the data in this way was part of developing the ThBA Version 02. Table 8-6 shows what columns were removed from or added to the hierarchical structure of ThBA Version 01 (Table 8-3). Some columns, however, remained unchanged.

Table 8-7 shows how the ThBA Version 02 was reorganised for a part of the ThBA (passive strategies for 'increasing heat gain' in animals). Keeping this in mind and knowing that cooling, heating or a combination of these are the thermal challenges of the four case studies (Table 8-1), it was decided to reorganise the whole of ThBA Version 01 to become Version 02 (Table 8-8).

When 'heating' was the thermal challenge, the relevant actions were 'generating heat', 'preventing heat loss', 'stealing heat', 'decreasing heat loss' and 'increasing heat gain'. The biological solutions related to animals and plants needed to be amalgamated to make the ThBA Version 02 under the new categorisation system developed earlier (Table 8-7). Solutions in the ThBA Version 02 appeared in the same order as those in Version 01.

Table 8-8 shows the actions are scattered. It was therefore decided to group the same actions to facilitate a comparison of the solutions documented for each (Table 8-9). This led to the creation of the ThBA Version 03 for heating challenges.

Having grouped the relevant actions, it was observed that some had similar heat transfer variables as shown in the 'variable in the equation' column. Sorting the relevant actions based on the main parameter in their heat transfer equation seemed to assist recognition of possible design strategies, as the equations governing homeostasis in both buildings and living organisms are the same. Table 8-10 was the final reorganisation of the relevant actions when the thermal challenge was 'heating' (the ThBA Version 04 for heating challenges). Using the same process, the ThBA version 04 for cooling challenges was created and contained the three actions of 'decreasing heat gain', 'avoiding heat gain' and 'increasing heat loss'. Table 8-11 shows the ThBA Version 04 for cooling challenges.

Table 8-6. Added, removed and unchanged columns in Table 8-3.

	Added	Removed	Remained	Merged into 'Solution'				Removed	Removed		Added	Added	Added
		A	B	C	D	E		F	G	H			
		Parent action	Action	Strategy	Type	Sub-type		Means	Organism	Link	Heat transfer method	Variable in the equation	Temporary/permanent
A/P	PA/AC											Changed Links for the main thermoregulation principles		

Table 8-7. Part of the reorganised ThBA Version 02 using data from the ThBA Version 01: passive strategies of 'increasing heat gain' for animals.

Animals/ Plants	Passive/ Active	Action	Solution	Links for the main thermoregulation principles		Heat transfer method	Variable in the equation	Temporary/ Permanently
				Column 4	Column 2			
A/P	PA/AC	What change in the heat exchange mechanism maintains homeostasis?	What is the solution?	What principle is central to the solution?	What properties of a living organism do the central thermal adaptation principles belong to?	What method of heat transfer allows for thermoregulation?	Which parameter in the heat transfer equation is changed?	TE/PE
A	PA	Increasing heat gain	Nest building					
A	PA	Increasing heat gain	Orientation					
A	PA	Increasing heat gain	Shuttling	Surface area	Morphological properties	Convective heat transfer	Surface area	TE
A		Increasing heat gain	Thermal gradient driven					
A	PA	Increasing heat gain	Heliothermy/ basking					
A	PA	Increasing heat gain	Thigmothermy					

Table 8-8. ThBA Version 02: scattered actions related to 'heating' as the thermal challenge.

Passive/active	Animals/plants	Action	Solution	Links for the main thermoregulation principles		Method	Variable	Temporary/Permanent
				Column 4	Column 2			
PA/AC	A/p	What change in the heat exchange mechanism maintains homeostasis?	What is the solution?	What principle is central to the solution?	What properties of a living organism do the central thermal adaptation principles belong to?	What method of heat transfer allows for thermoregulation?	What is the main parameter in the heat transfer equation?	TE/PE
AC	A	Heat generation	Clustering/huddling/nesting	Population	Chemical properties	-	Metabolism	T
AC	A	Heat generation	Non-shivering thermogenesis	Fat burning	Chemical properties	-	Metabolism	T
AC	A	Heat generation	Shivering thermogenesis	Muscle activity	Chemical properties	-	Metabolism	T
Ac	P	Heat generation	Metabolic heat generation	Electron transport	Chemical properties	-	Metabolism	T
PA	A	Stealing heat	Kleptothermy	Distance	Spatial properties	Conductive heat gain	Temperature gradient	T
PA	A	Increasing heat gain	Nest building	Orientation	Spatial properties	Solar heat gain	Surface area	P
PA	A	Increasing heat gain	Orientation	Orientation	Spatial properties	Solar heat gain	Surface area	T
PA	A	Increasing heat gain	Shuttling	Location	Spatial properties	Convective heat gain	Temperature gradient	T

Table 8-8 Contd. ...

...Table 8-8 Contd.

Passive/active	Animals/plants	Action	Solution	Links for the main thermoregulation principles		Method	Variable	Temporary/Permanent
				Column 4	Column 2			
PA	A	Increasing heat gain	Movement along thermal gradient	Distance	Spatial properties	Convective heat gain	Temperature gradient	T
PA	A	Increasing heat gain	Heliothermy/basking	Surface area	Morphological properties	Solar heat gain	Surface area	T
PA	A	Increasing heat gain	Thigmothermy	Distance	Spatial properties	Conductive heat gain	Temperature gradient	T
PA	A	Decreasing heat loss	Hunch	Surface area	Morphological properties	Convective heat loss	Surface area	T
PA	A	Decreasing heat loss	Back rest	Surface area	Morphological properties	Convective heat loss	Surface area	T
PA	A	Decreasing heat loss	Permanent insulation	Thermal conductivity	Material properties	Conductive heat loss	Heat transfer coefficient	P
PA	A	Decreasing heat loss	Nest material	Thermal conductivity	Material properties	Conductive heat loss	Heat transfer coefficient	P
PA	A	Avoiding heat loss	Burrowing	Location	Spatial properties	Convective heat loss	Temperature gradient	T
PA	A	Avoiding heat loss	Migration	Location	Spatial properties	Convective heat loss	Temperature gradient	T
PA	A	Avoiding heat loss	Daily torpor	Metabolism	Energy use (chemical properties)	Convective heat loss	Heart rate (drops)	T
PA	A	Avoiding heat loss	Estivation	Metabolism	Energy use (chemical properties)	Evaporative heat loss	Heart rate (drops)	T

PA	A	Avoiding heat loss	Hibernation	Metabolism	Energy use (chemical properties)	Convective heat loss	Heart rate (drops)	T
AC	A	Decreasing heat loss	Temporary growth of coats	Thermal conductivity	Material properties	Conductive heat loss	Heat transfer coefficient	P
AC	A	Decreasing heat loss	Waterproofing	Surface area	Substance properties	Evaporative heat loss	Surface area	T
AC	A	Decreasing heat loss	Preening	Water flow	Substance properties	Evaporative heat loss	surface area	T
AC	A	Decreasing heat loss	Piloerection	Thickness	Morphological properties	Conductive heat loss	Heat transfer coefficient	T
AC	A	Decreasing heat loss	Ptiloerection	Thickness	Morphological properties	Conductive heat loss	Heat transfer coefficient	T
AC	A	Decreasing heat loss	Hidromeiosis	Surface area	Morphological properties	Evaporative heat loss	Surface area	T
AC	A	Decreasing heat loss	Body shrinkage	Surface area	Morphological properties	Evaporative heat loss	Surface area	T
AC	A	Decreasing heat loss	Countercurrent heat exchange	Distance	Spatial properties	Convective heat loss	Temperature gradient	P
AC	A	Decreasing heat loss	Vasoconstriction	Diameter, fluid flow rate	Morphological properties	Convective heat loss	Heat transfer coefficient	T
AC	A	Avoiding heat loss	Supercooling	Freezing point	Substance properties	-	Heat transfer (stop)	T
AC	A	Avoiding heat loss	Extracellular freezing	Water flow	Substance properties	-	Heat transfer (stop)	T
PA	P	Increasing heat gain	Orientation	Orientation	Spatial properties	Solar heat gain	Surface area	T

Table 8-8 Contd. ...

...Table 8-8 Contd.

Passive/active	Animals/plants	Action	Solution	Links for the main thermoregulation principles		Method	Variable	Temporary/Permanent
				Column 4	Column 2			
AC	P	Increasing heat gain	Cuticle thickness control	Thickness	Morphological properties	Solar heat gain	Heat transfer coefficient	P
PA	P	Increasing heat gain	Pigmentation	Colour	Material properties	Solar heat gain	Solar heat gain coefficient (absorption)	P
AC	P	Increasing heat gain	Palisade cells' length and intensity	Depth	Morphological properties	Solar heat gain	Solar heat gain coefficient (transmission)	P
AC	P	Increasing heat gain	Accumulation response	Location, surface area	Spatial properties	Solar heat gain	Solar heat gain coefficient (absorption)	T
PA	P	Increasing heat gain	Nyctinastic movement	Surface area	Morphological properties	Solar heat gain	Surface area	T
PA	P	Increasing heat gain	Heliotropism	Orientation	Spatial properties	Solar heat gain	Surface area	T
PA	P	Increasing heat gain	Growth forms	Distance	Spatial properties	Convective heat gain Radiative heat gain	Temperature gradient	
PA	P	Increasing heat gain	Lens-shaped epidermal cells	Shape	Morphological properties	Solar heat gain	Solar heat gain coefficient (absorption)	P
PA	P	Decreasing heat loss	Leaf size control	Surface area	Morphological properties	Evaporative heat loss	Surface area	T

PA	P	Decreasing heat loss	Growth forms	Surface area	Morphological properties	Evaporative heat loss	Surface area	P
AC	P	Decreasing heat loss	Stomata closure	Surface area	Morphological properties	Evaporative heat loss	Surface area	T
AC	P	Decreasing heat loss	Decreased stomata density	Density	Morphological properties	Evaporative heat loss	Surface area	T
PA	P	Decreasing heat loss	Nyctinastic movement	Surface area	Morphological properties	Convective heat loss	Surface area	T
AC	P	Avoiding heat loss	Supercooling	Freezing point	Substance properties	-	Heat transfer (stop)	T
AC	P	Avoiding heat loss	Extracellular freezing	Water flow	Substance properties	-	Heat transfer (stop)	T
AC	P	Decreasing heat loss	Photosynthesis	Photosynthesis rate	-	-	-	T
AC	P	Decreasing heat loss	Waxy cuticle	Water flow	Substance properties	Evaporative heat loss	Surface area	P
PA	P	Decreasing heat loss	Epidermal hair or trichomes	Airflow	Material properties	Evaporative heat loss	Wind speed	P
AC	P	Decreasing heat loss	Losing leaves	Surface area	Morphological properties	Evaporative heat loss	Surface area	T
PA	P	Decreasing heat loss	Leaf rolling	Surface area	Morphological properties	Evaporative heat loss	Surface area	T
AC	P	Decreasing heat loss	Sunken stomata	Airflow	Material properties	Evaporative heat loss	Wind speed	P

Table 8-9. ThBA Version 03: grouping similar actions related to 'heating' as the thermal challenge.

Passive/active	Animals/plants	Action	Solution	Links for the main thermoregulation principles		Method	Variable	Temporary/Permanent
				Column 4	Column 2			
PA/AC	A/p	What change in the heat exchange mechanism maintains homeostasis?	What is the solution?	What principle is central to the solution?	What properties of a living organism do the central thermal adaptation principles belong to?	What method of heat transfer allows for thermoregulation?	What is the main parameter in the heat transfer equation?	TE/PE
PA	A	Heat generation	Clustering/huddling/nesting	Population	Chemical properties	-	Metabolism	T
AC	A	Heat generation	Non-shivering thermogenesis	Fat burning	Chemical properties	-	Metabolism	T
AC	A	Heat generation	Shivering thermogenesis	Muscle activity	Chemical properties	-	Metabolism	T
AC	P	Heat generation	Metabolic heat generation	Electron transport	Chemical properties	-	Metabolism	T
PA	A	Stealing heat	Kleptothermy	Distance	Spatial properties	Conductive heat gain	Temperature gradient	T
PA	A	Increasing heat gain	Nest building	Orientation	Spatial properties	Solar heat gain	Surface area	P
PA	A	Increasing heat gain	Orientation	Orientation	Spatial properties	Solar heat gain	Surface area	T
PA	A	Increasing heat gain	Shuttling	Location	Spatial properties	Convective heat gain	Temperature gradient	T

PA	A	Increasing heat gain	Movement along thermal gradient	Distance	Spatial properties	Convective heat gain	Temperature gradient	T
PA	A	Increasing heat gain	Heliothermy/basking	Surface area	Morphological properties	Solar heat gain	Surface area	T
PA	A	Increasing heat gain	Thigmothermy	Distance	Spatial properties	Conductive heat gain	Temperature gradient	T
PA	P	Increasing heat gain	Orientation	Orientation	Spatial properties	Solar heat gain	Surface area	T
AC	P	Increasing heat gain	Cuticle thickness control	Thickness	Morphological properties	Solar heat gain	Heat transfer coefficient	P
PA	P	Increasing heat gain	Pigmentation	Colour	Material properties	Solar heat gain	Solar heat gain coefficient (absorption)	P
AC	P	Increasing heat gain	Palisade cells length and intensity	Depth	Morphological properties	Solar heat gain	Solar heat gain coefficient (transmission)	P
AC	P	Increasing heat gain	Accumulation response	Location, surface area	Spatial properties	Solar heat gain	Solar heat gain coefficient (absorption)	T
PA	P	Increasing heat gain	Nyctinastic movement	Surface area	Morphological properties	Solar heat gain	Surface area	T
PA	P	Increasing heat gain	Heliotropism	Orientation	Spatial properties	Solar heat gain	Surface area	T
PA	P	Increasing heat gain	Growth forms	Distance	Spatial properties	Convective heat gain Radiative heat gain	Temperature gradient	

Table 8-9 Contd. ....

...Table 8-9 Contd.

Passive/ active	Animals/ plants	Action	Solution	Links for the main thermoregulation principles				Method	Variable	Temporary/ Permanent
				Column 4	Column 2					
PA	P	Increasing heat gain	Lens-shaped epidermal cells	Shape	Morphological properties			Solar heat gain	Solar heat gain coefficient (absorption)	P
PA	A	Decreasing heat loss	Hunch	Surface area	Morphological properties			Convective heat loss	Surface area	T
PA	A	Decreasing heat loss	Back rest	Surface area	Morphological properties			Convective heat loss	Surface area	T
PA	A	Decreasing heat loss	Permanent insulation	Thermal conductivity	Material properties			Conductive heat loss	Heat transfer coefficient	P
PA	A	Decreasing heat loss	Nest building	Thermal conductivity	Material properties			Conductive heat loss	Heat transfer coefficient	P
AC	A	Decreasing heat loss	Temporary growth of coats	Thermal conductivity	Material properties			Conductive heat loss	Heat transfer coefficient	P
AC	A	Decreasing heat loss	Waterproofing	Surface area	Substance properties			Evaporative heat loss	Surface area	T
AC	A	Decreasing heat loss	Preening	Water flow	Substance properties			Evaporative heat loss	surface area	T
AC	A	Decreasing heat loss	Piloerection	Thickness	Morphological properties			Conductive heat loss	Heat transfer coefficient	T
AC	A	Decreasing heat loss	Ptiloerection	Thickness	Morphological properties			Conductive heat loss	Heat transfer coefficient	T
AC	A	Decreasing heat loss	Hidromeiosis	Surface area	Morphological properties			Evaporative heat loss	Surface area	T

AC	A	Decreasing heat loss	Body shrinkage	Surface area	Morphological properties	Evaporative heat loss	Surface area	T
AC	A	Decreasing heat loss	Countercurrent heat exchange	Distance	Spatial properties	Convective heat loss	Temperature gradient	P
AC	A	Decreasing heat loss	Vasoconstriction	Diameter, fluid flow rate	Morphological properties	Convective heat loss	Heat transfer coefficient	T
PA	P	Decreasing heat loss	Leaf size control	Surface area	Morphological properties	Convective heat loss	Surface area	T
PA	P	Decreasing heat loss	Growth forms	Surface area	Morphological properties	Evaporative heat loss	Surface area	P
AC	P	Decreasing heat loss	Stomata closure	Surface area	Morphological properties	Evaporative heat loss	Surface area	T
AC	P	Decreasing heat loss	Decreased stomata density	Density	Morphological properties	Evaporative heat loss	Surface area	T
PA	P	Decreasing heat loss	Nyctinastic movement	Surface area	Morphological properties	Convective heat loss	Surface area	T
AC	P	Decreasing heat loss	Photosynthesis	Photosynthesis rate	-	-	-	T
AC	P	Decreasing heat loss	Waxy cuticle	Water flow	Substance properties	Evaporative heat loss	Surface area	P
PA	P	Decreasing heat loss	Epidermal hair or trichomes	Airflow	Material properties	Evaporative heat loss	Wind speed	P
AC	P	Decreasing heat loss	Losing leaves	Surface area	Morphological properties	Evaporative heat loss	Surface area	T
PA	P	Decreasing heat loss	Leaf rolling	Surface area	Morphological properties	Evaporative heat loss	Surface area	T
AC	P	Decreasing heat loss	Sunken stomata	Airflow	Material properties	Evaporative heat loss	Wind speed	P

Table 8-9 Contd. ...

...Table 8-9 Contd.

Passive/ active	Animals/ plants	Action	Solution	Links for the main thermoregulation principles		Method	Variable	Temporary/ Permanent
				Column 4	Column 2			
PA	A	Avoiding heat loss	Burrowing	Location	Spatial properties	Convective heat loss	Temperature gradient	T
PA	A	Avoiding heat loss	Migration	Location	Spatial properties	Convective heat loss	Temperature gradient	T
PA	A	Avoiding heat loss	Daily torpor	Metabolism	Energy use (chemical properties)	Convective heat loss	Heart rate (drops)	T
PA	A	Avoiding heat loss	Estivation	Metabolism	Energy use (chemical properties)	Evaporative heat loss	Heart rate (drops)	T
PA	A	Avoiding heat loss	Hibernation	Metabolism	Energy use (chemical properties)	Convective heat loss	Heart rate (drops)	T
AC	A	Avoiding heat loss	Supercooling	Freezing point	Substance properties	-	Heat transfer (stop)	T
AC	A	Avoiding heat loss	Extracellular freezing	Water flow	Substance properties	-	Heat transfer (stop)	T
AC	P	Avoiding heat loss	Supercooling	Freezing point	Substance properties	-	Heat transfer (stop)	T
AC	P	Avoiding heat loss	Extracellular freezing	Water flow	Substance properties	-	Heat transfer (stop)	T

Table 8-10. ThBA Version 04: the final reorganisation of relevant actions related to 'heating' as the thermal challenge.

Passive/ active	Animals/ plants	Action	Solution	Links for the main thermoregulation principles		Method	Variable	Temporary/ Permanent
				Column 4	Column 2			
PA/AC	A/p	What change in the heat exchange mechanism maintains homeostasis?	What is the solution?	What principle is central to the solution?	What properties of a living organism do the central thermal adaptation principles belong to?	What method of heat transfer allows for thermoregulation?	What is the main parameter in the heat transfer equation?	TE/PE
AC	A	Heat generation	Clustering/ huddling/ nesting	Population	Chemical properties	-	Metabolism	T
AC	A	Heat generation	Non-shivering thermogenesis	Fat burning	Chemical properties	-	Metabolism	T
AC	A	Heat generation	Shivering thermogenesis	Muscle activity	Chemical properties	-	Metabolism	T
AC	P	Heat generation	Metabolic heat generation	Electron transport	Chemical properties	-	Metabolism	T
PA	A	Stealing heat	Kleptothermy	Distance	Spatial properties	Conductive heat gain	Temperature gradient	T
PA	A	Increasing heat gain	Shuttling	Location	Spatial properties	Convective heat gain	Temperature gradient	T
PA	A	Increasing heat gain	Movement along thermal gradient	Distance	Spatial properties	Convective heat gain	Temperature gradient	T
PA	A	Increasing heat gain	Thigmothermy	Distance	Spatial properties	Conductive heat gain	Temperature gradient	T

Table 8-10 Contd. ...

...*Table 8-10 Contd.*

Passive/ active	Animals/ plants	Action	Solution	Links for the main thermoregulation principles		Method	Variable	Temporary/ Permanent
				Column 4	Column 2			
PA	P	Increasing heat gain	Growth forms	Distance	Spatial properties	Convective heat gain Radiative heat gain	Temperature gradient	
PA	P	Increasing heat gain	Pigmentation	Colour	Material properties	Solar heat gain	Solar heat gain coefficient (absorption)	P
AC	P	Increasing heat gain	Accumulation response	Location, surface area	Spatial properties	Solar heat gain	Solar heat gain coefficient (absorption)	T
PA	P	Increasing heat gain	Lens-shaped epidermal cells	Shape	Morphological properties	Solar heat gain	Solar heat gain coefficient (absorption)	P
AC	P	Increasing heat gain	Palisade cells' length and intensity	Depth	Morphological properties	Solar heat gain	Solar heat gain coefficient (transmission)	P
PA	A	Increasing heat gain	Nest building	Orientation	Spatial properties	Solar heat gain	Surface area	P
PA	A	Increasing heat gain	Orientation	Orientation	Spatial properties	Solar heat gain	Surface area	T
PA	A	Increasing heat gain	Heliothermy/ basking	Surface area	Morphological properties	Solar heat gain	Surface area	T
PA	P	Increasing heat gain	Orientation	Orientation	Spatial properties	Solar heat gain	Surface area	T
PA	P	Increasing heat gain	Nyctinastic movement	Surface area	Morphological properties	Solar heat gain	Surface area	T

PA	P	Increasing heat gain	Heliotropism	Orientation	Spatial properties	Solar heat gain	Surface area	T
PA	P	Increasing heat gain	Cuticle thickness control	Thickness	Morphological properties	Solar heat gain	Heat transfer coefficient	P
PA	A	Decreasing heat loss	Hunch	Surface area	Morphological properties	Convective heat loss	Surface area	T
PA	A	Decreasing heat loss	Back rest	Surface area	Morphological properties	Convective heat loss	Surface area	T
AC	A	Decreasing heat loss	Waterproof skin	Surface area	Substance properties	Evaporative heat loss	Surface area	T
AC	A	Decreasing heat loss	Preening	Water flow	Substance properties	Evaporative heat loss	Surface area	T
AC	A	Decreasing heat loss	Hidromeiosis	Surface area	Morphological properties	Evaporative heat loss	Surface area	T
AC	A	Decreasing heat loss	Body shrinkage	Surface area	Morphological properties	Evaporative heat loss	Surface area	T
PA	P	Decreasing heat loss	Leaf size control	Surface area	Morphological properties	Evaporative heat loss	Surface area	T
PA	P	Decreasing heat loss	Growth forms	Surface area	Morphological properties	Evaporative heat loss	Surface area	P
AC	P	Decreasing heat loss	Stomata closure	Surface area	Morphological properties	Evaporative heat loss	Surface area	T
AC	P	Decreasing heat loss	Decreased stomata density and size	Density, size	Morphological properties	Evaporative heat loss	Surface area	T
PA	P	Decreasing heat loss	Nyctinastic movement	Surface area	Morphological properties	Convective heat loss	Surface area	T
AC	P	Decreasing heat loss	Waxy cuticle	Water flow	Substance properties	Evaporative heat loss	Surface area	P

Table 8-10 Contd. ...

...Table 8-10 Contd.

Passive/ active	Animals/ plants	Action	Solution	Links for the main thermoregulation principles		Method	Variable	Temporary/ Permanent
				Column 4	Column 2			
AC	P	Decreasing heat loss	Losing leaves	Surface area	Morphological properties	Evaporative heat loss	Surface area	T
PA	P	Decreasing heat loss	Leaf rolling	Surface area	Morphological properties	Evaporative heat loss	Surface area	T
PA	A	Decreasing heat loss	Permanent insulation	Thermal conductivity	Material properties	Conductive heat loss	Heat transfer coefficient	P
PA	A	Decreasing heat loss	Nest building	Thermal conductivity	Material properties	Conductive heat loss	Heat transfer coefficient	P
AC	A	Decreasing heat loss	Temporary growth of coats	Thermal conductivity	Material properties	Conductive heat loss	Heat transfer coefficient	P
AC	A	Decreasing heat loss	Piloerection	Thickness	Morphological properties	Conductive heat loss	Heat transfer coefficient	T
AC	A	Decreasing heat loss	Ptiloerection	Thickness	Morphological properties	Conductive heat loss	Heat transfer coefficient	T
AC	A	Decreasing heat loss	Vasoconstriction	Diameter, fluid flow rate	Morphological properties	Convective heat loss	Heat transfer coefficient	T
AC	A	Decreasing heat loss	Countercurrent heat exchange	Distance	Spatial properties	Convective heat loss	Temperature gradient	P
AC	P	Decreasing heat loss	Photosynthesis	Photosynthesis rate	-	-	-	T
PA	P	Decreasing heat loss	Epidermal hair or trichomes	Airflow	Material properties	Evaporative heat loss	Wind speed	P
AC	P	Decreasing heat loss	Sunken stomata	Airflow	Material properties	Evaporative heat loss	Wind speed	P

PA	A	Avoiding heat loss	Burrowing	Location	Spatial properties	Convective heat loss	Temperature gradient	T
PA	A	Avoiding heat loss	Migration	Location	Spatial properties	Convective heat loss	Temperature gradient	T
AC	A	Avoiding heat loss	Supercooling	Freezing point	Substance properties	-	Heat transfer (stop)	T
AC	A	Avoiding heat loss	Extracellular freezing	Water flow	Substance properties	-	Heat transfer (stop)	T
AC	P	Avoiding heat loss	Supercooling	Freezing point	Substance properties	-	Heat transfer (stop)	T
AC	P	Avoiding heat loss	Extracellular freezing	Water flow	Substance properties	-	Heat transfer (stop)	T
PA	A	Avoiding heat loss	Daily torpor	Metabolism	Energy use (chemical properties)	Convective heat loss	Heart rate (drops)	T
PA	A	Avoiding heat loss	Estivation	Metabolism	Energy use (chemical properties)	Evaporative heat loss	Heart rate (drops)	T
PA	A	Avoiding heat loss	Hibernation	Metabolism	Energy use (chemical properties)	Convective heat loss	Heart rate (drops)	T

Table 8-11. ThBA Version 04: relevant actions related to 'cooling' as the thermal challenge.

Passive/active PA/AC	Animals/plants A/p	Action: What change in the heat exchange mechanism maintains homeostasis?	Solution: What is the solution?	Links for the main thermoregulation principles — Column 4: What principle is central to the solution?	Column 2: What properties of a living organism do the central thermal adaptation principles belong to?	Method: What method of heat transfer allows for thermoregulation?	Variable: What is the main parameter in the heat transfer equation?	Temporary/Permanent TE/PE
PA	A	Decreasing heat gain	Orientation	Orientation	Spatial properties	Solar heat gain	Surface area	T
PA	P	Decreasing heat gain	Orientation	Orientation	Spatial properties	Solar heat gain	Surface area	T
PA	P	Decreasing heat gain	Nyctinastic movement (closing behaviour)	Surface area	Morphological properties	Solar heat gain	Surface area	T
PA	P	Decreasing heat gain	Growth forms	Surface area	Morphological properties	Solar heat gain	Surface area	P
PA	P	Decreasing heat gain	Trichomes/pubescence	Surface area	Morphological properties	Solar heat gain	Surface area	P
PA	P	Decreasing heat gain	Spines	Surface area	Morphological properties	Solar heat gain	Surface area	P
AC	P	Decreasing heat gain	Thylakoids structure	Surface area	Morphological properties	Solar heat gain	Surface area	P
PA	A	Decreasing heat gain	Sidewinding	Surface area	Morphological properties	Conductive heat gain	Surface area	T

				Surface area	Morphological properties	Conductive heat gain	Surface area	
PA	A	Decreasing heat gain	Stilting	Surface area	Morphological properties	Conductive heat gain	Surface area	T
PA	A	Decreasing heat gain	Nest building	Orientation	Spatial properties	Solar heat gain	Surface area	P
PA	A	Decreasing heat gain	Shuttling	Location	Spatial properties	Convective heat gain	Temperature gradient	T
PA	A	Decreasing heat gain	Movement along thermal gradient	Distance	Spatial properties	Convective heat gain	Temperature gradient	T
PA	A	Decreasing heat gain	Climbing	Distance	Spatial properties	Conductive heat gain	Temperature gradient	T
PA	A	Decreasing heat gain	Colour reflectance	Colour	Material properties	Solar heat gain	Solar heat gain coefficient (absorption)	T
PA	P	Decreasing heat gain	Pigmentation	Colour	Material properties	Solar heat gain	Solar heat gain coefficient (absorption)	P
PA	P	Decreasing heat gain	Light coloured hair	Colour	Material properties	Solar heat gain	Solar heat gain coefficient (absorption)	P
AC	P	Decreasing heat gain	Cuticle thickness control	Surface area	Morphological properties	Solar heat gain	Solar heat gain coefficient (absorption)	P
AC	P	Decreasing heat gain	Avoidance response	Location, surface area	Spatial properties	Solar heat gain	Solar heat gain coefficient (absorption)	T
AC	P	Decreasing heat gain	Palisade cells length and intensity	Depth	Morphological properties	Solar heat gain	Solar heat gain coefficient (transmission)	P

Table 8-11 Contd. ...

...Table 8-11 Contd.

Passive/active	Animals/plants	Action	Solution	Links for the main thermoregulation principles Column 4	Column 2	Method	Variable	Temporary/Permanent
PA	A	Decreasing heat gain	Permanent Insulation	Thermal conductivity	Material properties	Conductive heat gain	Heat transfer coefficient	P
AC	A	Decreasing heat gain	Temporary growth of coats	Thermal conductivity	Material properties	Conductive heat gain	Heat transfer coefficient	T
PA	A	Avoiding heat gain	Shading	Surface area	Morphological properties	Solar heat gain	Surface area	T
PA	A	Avoiding heat gain	Burrowing	Location	Spatial properties	Convective heat loss	Temperature gradient	T
PA	A	Avoiding heat gain	Migration	Location	Spatial properties	Convective heat loss	Temperature gradient	T
PA	A	Increasing heat loss	Climbing	Distance	Spatial properties	Conductive heat gain	Temperature gradient	T
PA	A	Increasing heat loss	Nest building	Mound structure, cavity shape	Morphological properties	Convective heat loss	Temperature gradient	P
AC	A	Increasing heat loss	Selective brain cooling	Diameter, fluid flow rate	Morphological properties	Convective heat loss	Heat transfer coefficient	T
AC	A	Increasing heat loss	Menopausal hot flush	Diameter, fluid flow rate	Morphological properties	Convective heat loss	Heat transfer coefficient	T
PA	A	Increasing heat loss	Saliva spreading	Surface area	Morphological properties	Evaporative heat loss	Surface area	T
AC	A	Increasing heat loss	Wallowing	Surface area	Morphological properties	Evaporative heat loss	Surface area	T

PA	P	Increasing heat loss	Stem elongated cells	Surface area	Morphological properties	Evaporative heat loss	Surface area	P
PA	P	Increasing heat loss	Growth forms	Surface area	Morphological properties	Evaporative heat loss	Surface area	P
AC	P	Increasing heat loss	Stomata opening	Surface area	Morphological properties	Evaporative heat loss	Surface area	T
AC	P	Increasing heat loss	Increased stomata density and size	Density, size	Morphological properties	Evaporative heat loss	Surface area	T
PA	P	Increasing heat loss	Leaf size control	Surface area	Morphological properties	Evaporative heat loss	Surface area	T
PA	P	Increasing heat loss	Nyctinastic movement (opening behaviour)	Surface area	Morphological properties	Convective heat loss	Surface area	T
AC	A	Increasing heat loss	Sweating	Surface area	Morphological properties	Evaporative heat loss	Surface area	T
AC	A	Increasing heat loss	Thermal hyperpnea	Volume	Morphological properties	Evaporative heat loss	Surface area	T
AC	A	Increasing heat loss	Thermal tachypnea	Volume	Morphological properties	Evaporative heat loss	Surface area	T
AC	A	Increasing heat loss	Gular fluttering	Surface area	Morphological properties	Evaporative heat loss	Surface area	T

8.5 Building A in Dunedin (using the ThBA Version 04, Test 01)

As Table 8-1 shows, the thermal challenge for Building A in Dunedin was heating; therefore, relevant solutions were found using the new version of the ThBA related to 'heating' (Table 8-10). As discussed earlier, the column named 'Links' shows the generalised thermal adaptation principle behind thermoregulation. The columns labelled 'Method' and 'Variable' show the heat transfer methods and the main parameters in their equations respectively.

The following sections outline solutions which were considered either irrelevant to Building A, or impossible to translate into architecture due to the current state of technology. The inappropriate solutions were identified by looking at the 'Variable' column, where strategies with similar main parameters in their heat transfer equations were grouped in one category within a box with a thick line boundary (Table 8-10). The appropriate solutions were highlighted in grey.

8.5.1 Inappropriate solutions

As the 'Variable' column in Table 8-10 shows, some solutions related to the five main actions (heat generation, stealing heat, increasing heat gain, decreasing heat loss and avoiding heat loss) seemed inappropriate for redesigning Building A in Dunedin. As mentioned earlier, the solutions in Table 8-10 were categorised based on the main heat transfer parameter they had in common. This meant the inappropriate solutions were those for which the main heat transfer parameters were inappropriate for Building A in Dunedin.

The process started by looking at solutions with similar main parameters. In order to translate biological principles to architectural principles, the same architectural variable needed to be controlled to allow for parallel thermoregulation in the building design. Accordingly, the solutions were recognised as irrelevant if changing the main parameter of the heat transfer was not feasible for this particular case study.

8.5.1.1 Action one: increasing heat gain + generating heat + stealing heat

Population, fat burning, muscle activity, electron transport: In a nest, an increase in the population will increase its internal temperature. A parallel is thinking of users as the nest population which means one approach to raise the temperature inside a building is to fit more people in the same space for metabolic heat generation. While it seems an odd approach, it would appear to be effective in winter but would impose an extra cooling load in summer. To make this strategy useful, the users have to follow a seasonal work schedule. The nearest building type to use this strategy is the traditional house barn where animals and people were sheltered under the same roof. In winter, heat from the animals would heat the interior used by the farming family, while in summer the animals would be outside on the grass (Tishler and Witmer 1986). Such a strategy would not work for an office building unless some people were prepared to work outside in summer. Additionally, the number of users would need to be significantly large to meet the heating requirements of a building in winter, as roughly each person gives off 100 watts, so 10 people in a room are the equivalent of one bar of an electric fire.

For fat burning and muscle activity, the heat generation happens due to the chemical reactions in the cells, where food is changed into energy. As discussed before, for the former, the equivalent strategy in buildings would be running HVAC systems, which is not an energy-efficient way of designing buildings. For the latter, there is no equivalent strategy although renewable sources can be seen as a very weak and parallel to muscle activity. Similarly, there is no equivalent strategy for heat generation in plants happening either through electron transport or respiration.

Temperature gradient: As organisms repeatedly alter the distance between their bodies and a heat source to control body temperature, in buildings this could happen through changing the distance between the zones to be heated and the main source of heat gain. One way of doing this would be to move Building A to a warmer climate, but this would not be feasible. Such a design strategy was also identified as irrelevant because Building A is a skin-load-dominated building in which the nature of the envelope causes the high energy use. Given this, the equivalent mechanism in architecture, which might be drawing zones to be heated closer to the envelope to gain heat from an external source like the sun, is not feasible.

Solar heat gain coefficient (absorption and transmission): Organisms increase solar heat gain through absorption or transmission by changing morphological properties, material properties or spatial properties. These actions happen through change in the shape and depth of light penetrating organs (morphological change), the colour of the tissues (material change) and a change in the location or the distance from a heat source (spatial change). For Building A in Dunedin, the only feasible solution is changing the colour of the building envelope. The use of colour for adjusting solar heat gain in buildings is not an innovative strategy. In fact, this has been used for a long time. For example, glasshouses have been painted white in summer to avoid overheating by reflecting more of the incoming solar radiation. However, the aim here has been to avoid overheating. A modern example of the effect of colour on solar heat gain is the theoretical research into how colour might change the performance of PV panels (Halme and

Mäkinen 2019), the thought being that coloured PV panels would make more attractive building facades. However, here such colouring led to a reduction in performance compared to the common black PV. This could be viewed as a perverse example of copying nature to reduce energy efficiency. A change in the distance from a heat source has been discussed above.

Putting an atrium into a building could be thought of as a morphological change in the building anatomy. An atrium can be seen as the architectural equivalent of the changes taking place in the light penetrating organs in the bodies of organisms. However, for an atrium the change takes place at the very different scale of the building envelope. Such a change might be feasible for an existing building undergoing radical refurbishment, but this was not the nature of the problem set for Building A in Dunedin as the atrium would lead to a loss of useable floor area. In a different approach, there might be change in the structure of materials used for windows at the nanoscale. An example of this is the recent development of transparent solar cells that look like tinted glass. Semi-transparent photovoltaic glazing has been recently linked to the design of energy-efficient buildings (Qiu and Yang 2020). Reglazing building A in this way could be thought of as a material change that would increase solar abruption, though here for the purpose of generating electricity rather than heat.

*Surface area: Termites increase solar heat gain by modifying the properties of their nest when building it, such as orienting the nest walls and openings towards the sun. Similarly, the key principle behind heliothermy, which is the orientation of leaves and animals towards the sun, is to expose a larger surface area of an organ or tissue to direct solar radiation. As discussed earlier, an equivalent strategy in building design could be the creation of a building envelope that follows the sun path for maximum solar gain. Likewise, plants change the sun-exposed surface area of their petals using the opening and closing behaviour.

The heat balance characteristics of Building A show there is conductive and radiative heat gain through the windows. This means that for Building A in Dunedin, a temporary increase in the surface area or reorientation of the windows could allow more solar gain in winter, which would consequently reduce the need for heating energy. This might be achieved through fitting windows with shutters which were only opened at certain times of the day and year to adjust the exposed area of the glazing. This is a strategy used at the domestic scale in many traditional dwellings, both to keep heat in at night and out during a hot sunny day. To use this strategy at a larger scale, control of the shutters might need to be done automatically for which energy is required. To evaluate if this is an energy-efficient solution, an energy balance equation would be needed to ensure the gains exceeded the extra energy needed for the mechanical control. There is evidence that giving people control, even in a shared office situation, can enhance environmental performance and productivity (Alt et al. 2015), so having control over opening shutters could be part of this.

8.5.1.2 *Action two: decreasing heat loss*

Surface area: Decreasing surface area to reduce evaporative heat loss is the central thermoregulatory principle in waterproof skin, hidromeiosis, body shrinkage, leaf size control, growth forms, stomata closure, decreased stomatal density, losing leaves, and leaf rolling. The two strategies of preening and waxy cuticle both control heat loss through adjusting water flow. However, none of these strategies are relevant for Building A in Dunedin.

In biological complementary strategies, responses to one stressor cause or are associated with regulating another stressor. For decreasing heat loss, except for hunching, back rest, growth forms and nyctinastic movement, regulating water flow is the main means of thermoregulation. These were all considered irrelevant to Building A in Dunedin as heat loss in the building does not happen due to water evaporation, but instead to temperature difference.

As discussed in 6.1.3, animals hunch and birds use back rest to reduce the surface area of their body in order to decrease conductive heat loss through their skin. Likewise, the movement of flower petals through closure decreases heat loss through convection by reducing the exposed surface area. As shown in Table 8-2, conductive heat loss through the opaque surfaces was high in Building A in Dunedin. Consequently, an equivalent design solution to these strategies would be reducing the surface area of the building envelope during winter, which is not possible.

Temperature gradient: As explained in 8.5.1.1, the strategy of changing the distance between the building and heat source to control the temperature gradient would not be feasible for Building A in Dunedin.

Wind speed: One variable affecting evaporative heat transfer is wind speed. The problem is that the evaporative heat loss is reduced significantly in very low wind speed. To translate the two biological solutions of epidermal hair and sunken stomata into architectural design principles, water would need to be the medium through which evaporative heat loss was controlled for regulating temperature. For Building A in Dunedin, although conductive heat loss through opaque surfaces was high, this was not associated with water flow, and this category of solutions was deemed irrelevant.

The link between thermoregulation and wind in organisms relates to changes in environmental conditions, rather than in the organisms themselves. Thinking of a building as an organism, its thermoregulation will be dependent on the external climatic condition. Given wind flow is intermittent, it cannot be fully trusted to be effective for building thermoregulation, whereas in nature it is beneficial as organisms intelligently switch strategies until their body reaches a comfortable temperature.

8.5.1.3 Action three: avoiding heat loss

Temperature gradient: As discussed in 8.5.1.1, controlling the temperature gradient by changing the location of the zones is not applicable to redesigning Building A in Dunedin.

Stop heat transfer/metabolism: The biological strategies in this category use a different process for thermal adaption as the heat transfer stops so organisms can survive harsh climates. Heat transfer stops when organisms cannot tolerate the severity of the extremes even if they decrease or increase heat loss significantly. For example, the freezing point of water in a cell reduces to allow the fluid in the cell to stay at a lower temperature. This means the freezing point of the fluid in the cell remains below the ambient temperature during cold seasons. Likewise, extracellular freezing happens through solidification of the cell wall to inhibit water loss. The crystallisation of the walls prevents water from flowing out of the cell. This will ultimately stop dehydration of the cells and hence, ensure the survival of the organism. Given this, a temporary pause in heat transfer enables survival.

There is a major difference between living organisms and buildings, as the former can stop their activities for a temporary period to save energy for the future when the period of extreme environmental conditions is passed. However, the temporary shutdown of office buildings on a regular basis is not feasible, though this does happen as a one-off event in extreme weather conditions when people are sent home early (Edge 2011). In the light of this, the solutions in this category were not applicable to an office building.

Drop heart rate/photosynthesis rate: Reducing metabolism and thus, heat generation in animals during torpor, estivation and hibernation is an avoidance mechanism that organisms use to acclimate to their environment. The analogy of reducing the heart rate in animals for Building A in Dunedin would be reducing energy use. This could be either reflected in turning off the HVAC systems or stopping the use of workspaces. Neither of these seemed a reasonable solution to avoiding heat loss in Building A in Dunedin.

8.5.2 Appropriate solutions

Solar heat gain coefficient (absorption and transmission): From all solutions related to this variable, the 'colour' strategy seems appropriate (see 8.5.1.1, solar heat gain coefficient).

Heat transfer coefficient for decreasing heat loss: Among all solutions from the ThBA in which the heat transfer coefficient was the main parameter in the heat transfer equation, some were relevant to Building A in Dunedin. Looking at column 4 of the 'Links' section in Table 8-10, except for vasoconstriction, the heat transfer coefficient changes due to an increase in either the thickness (morphological properties) or thermal conductivity (material properties) of an insulating layer in biological tissues. The change in the heat transfer equation limits conductive heat loss. In vasoconstriction, the heat transfer reduces due to the decrease in convective heat transfer from the blood vessels to the skin.

As some biological organisms control the thickness of the insulating layer, altering the thermal conductivity of the wall might be a solution for use in building design. A semi-similar approach for Building A in Dunedin would be decreasing the thermal conductivity of the insulating layer of the envelope but this needs to be temporary as the heat challenge occurs in winter. To make this solution work, the insulating layer needs to be movable. It might be easier to allow the users to adjust their clothing and wear more layers in winter than adjust the insulation in the building envelope. There are links between a more casual dress code and office productivity, not least because wearing a collar and tie can restrict the flow of blood to the brain (Lüddecke et al. 2018), which could reduce both motivation and productivity. However, this is a management rather than a design issue.

A more realistic solution would be to add more insulation to the opaque surfaces of Building A to reduce heat loss, which would also be beneficial in preventing heat gain through the envelope in summer, providing the building interior was well ventilated to avoid a build-up of heat. At this stage, it seemed pointless to redesign Building A, since insulation is not a novel technique for thermal regulation in buildings as it has been used by architects for centuries in many places around the world. The ThBA failed to come up with an innovative solution to the thermal challenge of heating.

8.6 Building A in Auckland (using the ThBA version 04, Test 02)

As explained in 8.3, to test the effectiveness of the ThBA for the climate extremes in New Zealand, it was necessary to search it again for Building A in Auckland.

Table 8-1 shows Building A in Auckland has the two thermal challenges of cooling and heating. As the heat transfer characteristics of this building in Auckland were similar to those in Dunedin (Table 8-2) there was no point in looking for solutions to the heating challenge as this had already been done for Building A in Dunedin. Therefore, the relevant part of the ThBA (Table 8-11) was searched for appropriate solutions. As for Table 8-10, strategies with similar main parameters in their heat transfer equations were grouped in one category within a box with a thick line boundary and appropriate solutions were highlighted in grey.

8.6.1 Action one: decreasing heat gain

8.6.1.1 Solar heat gain coefficient (absorption and transmission)

Using light colours for the external surfaces improves the thermal performance of buildings (Bansal et al. 1992, Synnefa et al. 2007) as the white-painted surfaces can reduce the peak cooling loads (Sadineni et al. 2011). Consequently, applying cool-coloured coatings in the manufacturing of building materials has been suggested for achieving energy savings (Synnefa et al. 2007). Figure 8-4 shows an example of using white-coloured roofs in Greece. Even though this was a relevant and feasible strategy for Building A in Auckland, it was only needed on a seasonal basis. While the permanent use of light-coloured coatings is a common sustainable design strategy, smart glass windows and intelligent colour-changing facades could be recognised as innovative design solutions that are expected to be incorporated into future buildings. Permanently tinted glass is already a common solution for reducing solar heat gain in buildings.

Figure 8-4. White-roofed Greek houses (Zwegers 2008).

For the two solutions of cuticle thickness control and avoidance response, solar heat gain is controlled through morphological changes taking place in the cuticle and chloroplast location for the first, while for palisade length and density, the reduced depth of light-channelling cells, which is a spatial property, decreases heat gain. Translation of either of these strategies into architectural principles for Building A is not feasible. Neither of the possible design solutions of either a temporary decrease in the surface area of the windows or inserting an atrium was feasible for Building A.

8.6.1.2 Heat transfer coefficient

This is discussed in 8.5.2.

8.6.2 Action two: avoiding heat gain and action three: increasing heat loss

No relevant solution was identified in these categories.

8.7 Architects know biomimicry by instinct

Reviewing the solutions for redesigning Building A in Auckland and Dunedin, it seems the simple translation of the majority of these solutions have been used in architectural practice and thus, the ThBA has not offered any new solutions to architects. The two columns of 'Method' and 'Variable' in Tables 8-10 and 8-11 seemed to play similar roles in biology and architecture and therefore, were used as a basis for categorising the in-use thermal adaptation strategies in building design. This categorisation might suggest another structure for the ThBA, which could include examples of organisms and 'Means' through which the architectural and thermal performance specifications of an office building would guide designers to find relevant solutions. Below are the suggested new categories.

8.7.1 Controlling conductive and convective heat gain through temperature gradient

Different building design approaches seem relevant to the idea of 'decreasing distance from a heat source' as a thermoregulatory principle related to heating.

Depending on the main source of heat gain in a building, the distance between the thermal zone and the heat source needs to be decreased to increase conductive or convective heat gain. For an internally-load-dominated office building the internal heat gains are the heat sources, while for a skin-load-dominated one, the aim would be to increase solar gain as the exterior heat source. For the former, an equivalent mechanism in architecture could be placing colder internal spaces next to those that generate too much heat, whether the heat is generated by the metabolic activities of users or gained through equipment and HVAC operations. For the latter, zones that need heat might be drawn closer to the envelope to gain heat from the outside. For buildings with significant potential for solar heat gain through the envelope, changing the arrangement of the thermal zones in a manner such that they share at least one surface with the envelope would increase heat gain.

8.7.2 *Controlling convective and conductive heat loss through temperature gradient*

The landscape surrounding a building creates microclimatic conditions for the interior and the external skin which could result in reduced energy consumption. Using vegetation for shade is a passive design strategy for protecting outdoor spaces and zones close to the envelope in summer, and for reducing wind speed and heat loss in winter.

Zero-energy earth-sheltered buildings seem to be inspired by the burrowing strategy (Vale and Vale 2013). Figure 8-5 shows a row of zero-energy south-facing houses with the northern side buried in the ground, and the ground mediating the temperature gradient between inside and outside.

Figure 8-5. Autonomous houses (Hockerton Housing project).

8.7.3 *Controlling solar heat gain through transmission and absorption*

The panel of biologists who evaluated the first version of the ThBA stated that in light-induced temperature regulation, there is an interaction between thermoregulatory stressors such that responding to light conditions induces the regulation of temperature. Similarly, increasing light absorption and transmission in a building would increase heat gain. Parallel building design solutions where change in the depth of light penetration through use of shape, structure and colour to increase light absorption and transmission, are almost conventional.

While, the morphological properties of palisade cells in plants adjust the light transmission so light can penetrate deep into leaves, windows and atria also transmit light into the building. Some plants employ the light-focusing shape of epidermal cells to increase heat gain and this might provide inspiration or architectural translation. Changing the colour of the façade has been commonly employed in buildings to control solar gain.

At the macroscale, there could be substantial light harvesting through the windows of a building by changing their shape and geometry. While the biological mechanism is permanent, any temporary architectural translation of such a permanent thermoregulatory mechanism in nature needs to be either manually or mechanically controlled. This biological solution could be translated into architectural design through the temporary geometric transformation of windows over the course of a year, although an energy balance check would need to be made.

At the microscale, light-responsive biomaterials incorporate an optical nanostructure to adjust light transmission. Hydrogels have been used as smart materials to control illumination through active and passive strategies. A new hydrogel biomaterial has recently been developed to be used in the building envelope. Windows made with hydrogels become opaque in response to high temperatures and hence inhibit light transmission by scattering light beams on the surface (Khoo and Shin 2018).

8.7.4 Controlling solar heat gain through surface area

This is discussed in 8.5.1.1.

8.7.5 Controlling evaporation through surface area

Passive and active evaporative cooling strategies are controlled by water surfaces or flows. There are several passive and active strategies used in conventional architecture such as green roofs/walls, cooled soil and evaporative coolers, although how the water is supplied needs to be part of the energy balance equation.

8.7.6 Controlling evaporation through air flow

The courtyard effect and solar chimneys have been used as passive evaporative cooling strategies. The cooled air from the surface of the water is drawn towards the warmer spaces as the warm air rises due to the stack effect. The best examples for application of these techniques can be seen in traditional Iranian houses in hot and dry climates, and in hot and humid climates.

8.7.7 Controlling conductive and convective heat gain through surface area

A similar approach in architecture to stilting and sidewinding, which are two strategies animals use to decrease conductive heat gain, could be stilt construction where a building is raised above the ground to encourage air flow around it in response to climatic conditions. Examples of such buildings are the stilt constructions and pile dwellings of southern Asia (Figure 8-6) that allow for natural ventilation in hot and humid climates (Ara and Rashid 2018).

Designing buildings in a compact form in a cold climate reduces the surface area of the building envelope and hence helps to minimise heat loss.

Figure 8-6. Raised construction (3coma14 2009).

8.7.8 Controlling convective and conductive heat loss and heat gain through heat transfer coefficient

Use of thermal insulation materials is one of the most popular sustainable design strategies, and units like R values ($m^2 \cdot K/W$) have been used to describe the thermal performance of insulation materials (Asdrubali et al. 2015). The historical background of insulation materials has been linked to the history of temporary dwellings made by prehistoric people (Bozsaky 2010). Because these buildings had to be moved and hence be light, the one strategy they could use to reduce heat loss was the use of fluffy materials, such as wool and other animal fibres, in the envelopes. A typical example would be the felt coverings of the Mongolian yurt (Figure 8-7).

Figure 8-7. A Mongolian Yurt (Vorel 2019).

References

3coma14. (2009). Inle-Yawnghwe. Retrieved from https://commons.wikimedia.org/wiki/ File:Inle-Yawnghwe.jpg, Creative Commons Attribution-Share Alike 3.0 Unported license: https://creativecommons.org/licenses/by-sa/3.0/deed.en.

Alt, E., Díez-de-Castro, E. P. and Lloréns-Montes, F. J. (2015). Linking employee stakeholders to environmental performance: The role of proactive environmental strategies and shared vision. *Journal of Business Ethics, 128*(1), 167–181.

Ara, D. R. and Rashid, M. (2018). An ethnic house form at the western margins of southeast Asia: the elusive south Asian stilt architecture of the Chittagong hill tracts. *The Asia Pacific Journal of Anthropology, 19*(1), 35–54.

Aran, A. (2007). Manufacturing properties of engineering materials—ITU Department of Mechanical Engineering Lecture Notes (PDF). Retrieved 20 July 2018 www2.isikun. edu.tr/personel/ahmet.aran/mfgprop.pdf.

Asdrubali, F., D'Alessandro, F. and Schiavoni, S. (2015). A review of unconventional sustainable building insulation materials. *Sustainable Materials and Technologies, 4*, 1–17.

Bansal, N., Garg, S. and Kothari, S. (1992). Effect of exterior surface colour on the thermal performance of buildings. *Building and Environment, 27*(1), 31–37.

Bozsaky, D. (2010). The historical development of thermal insulation materials. *Periodica Polytechnica Architecture, 41*(2), 49–56.

Edge, K. (2011). Storm closes 24 schools—Northland spared the worst. Retrieved 18 September 2018 https://www.nzherald.co.nz/northern-advocate/news/article. cfm?c_id=1503450&objectid=10972656.

Halme, J. and Mäkinen, P. (2019). Theoretical efficiency limits of ideal coloured opaque photovoltaics. *Energy & Environmental Science, 12*(4), 1274–1285.

Imani, N. (2020). *A Thermo-bio-architectural Framework (ThBA) for Finding Inspiration in Nature: Biomimetic Energy Efficient Building Design* (PhD thesis), Victoria University of Wellington, Wellington, New Zealand.

Imani, N. and Vale, B. (2020a). The development of a biomimetic design tool for building energy efficiency. *Biomimetics, 5*(4), 1–19. doi:https://doi.org/10.3390/biomimetics5040050.

Imani, N. and Vale, B. (2020b). A framework for finding inspiration in nature: Biomimetic energy efficient building design. *Energy and Buildings, 225*. doi:https://doi. org/10.1016/j.enbuild.2020.110296.

Khoo, C. K. and Shin, J.-W. (2018). *Designing with Biomaterials for Responsive Architecture.* Paper presented at the 36th International Conference on Education and Research in Computer Aided Design in Europe - eCAADe 2018, Lodz, Poland.

Lüddecke, R., Lindner, T., Forstenpointner, J., Baron, R., Jansen, O. and Gierthmühlen, J. (2018). Should you stop wearing neckties?—wearing a tight necktie reduces cerebral blood flow. *Neuroradiology, 60*(8), 861–864.

Qiu, C. and Yang, H. (2020). Daylighting and overall energy performance of a novel semi-transparent photovoltaic vacuum glazing in different climate zones. *Applied Energy, 276*, 115414.

Sadineni, S. B., Madala, S. and Boehm, R. F. (2011). Passive building energy savings: A review of building envelope components. *Renewable and Sustainable Energy Reviews, 15*(8), 3617–3631.

Synnefa, A., Santamouris, M. and Apostolakis, K. (2007). On the development, optical properties and thermal performance of cool colored coatings for the urban environment. *Solar Energy, 81*(4), 488–497.

Tishler, W. and Witmer, C. S. (1986). The housebarns of east-central Wisconsin. *Perspectives in Vernacular Architecture, 2*, 102–110.

Vale, B. and Vale, R. (2013). The Hockerton Housing Project, England. pp. 262–274. *In*: R. Vale and B. Vale (eds.). *Living Within a Fair Share Ecological Footprint.* New York, NY: Routledge.

Vorel, M. (2019). Mongolian Yurts. Retrieved from https://libreshot.com/mongolian-yurts/, Public domain license: https://creativecommons.org/licenses/publicdomain/.

Zwegers, A. (2008). Santorini, Oia. Retrieved from https://www.flickr.com/photos/azwegers/6246546927/in/photostream/, Creative Commons License: https://creativecommons.org/licenses/by/2.0/.

Chapter 9

Developing a Framework for Bio-Inspired Energy-Efficient Building Design

9.1 Introduction

The main drivers behind the development of the ThBA were the increasing interest in biomimetic architecture or bio-inspired building design and the ongoing debate on its promise to produce more sustainable buildings through aiding innovation (see 3.1.3). By global standards, buildings are a very significant consumer of energy, given that in 2018, 36% of global energy was used by buildings, while in the construction sector, 80% of this was being drawn from fossil fuels (IEA 2019). Therefore, the focus of the ThBA was on thermoregulation strategies found in the natural world.

In 1997, Beynus introduced the term 'biomimicry in architecture' and since then, people have attempted to develop a way to bridge the gap between biology and architecture with the goal of designing energy-efficient buildings. The recent attempt by Badarnah (2012) which was discussed in Chapter 4 (see 4.3.1) seemed to be neither comprehensive nor had been assessed by experts in biology. The ThBA as described in this book has tried to address both these issues. Perhaps more importantly, Badarnah's tool for accessing examples of thermoregulation on nature did not have an architectural side to aid the transition from what happens in organisms to what happens in a low-energy building. The ThBA addressed this shortcoming, making it possible to see if what occurs in nature could become an inspiration for new ways of designing low-energy buildings. However, because it was possible to test the ThBA, this raised the issue of how useful such a tool could truly be, since the relevant biological solutions found for the two thermally challenging case studies of Building A in Dunedin and Building A in Auckland as set out in

Chapter 8 through using the ThBA revealed that simple translations of the majority of solutions have already been used in building design. This discussion is the focus of this chapter.

9.2 The usefulness of the ThBA

Although the ThBA was only tested on a building with relatively simple thermal challenges, albeit those which were found in many buildings, almost all the suggestions raised by the ThBA have already been used in architectural design as a means of achieving energy efficiency. In a cold climate, the two routes to lowering energy demand are the inclusion of more insulation in the building envelope together with insulated mass to damp the swings in temperature. This approach was used in the BedZed housing in London (see Figure 2-2). In the animal kingdom, both mass and insulation, the latter in the form of fur, feathers and fat, have a place in controlling heat loss. Looking at buildings in hot climates, avoiding solar gain through use of a cool roof or even a white painted roof both have a direct parallel in nature in white and reflective flower petals. In a hot humid climate raising the building above the ground to allow for better air flow for cooling is again a strategy that can be found in the natural world in form of stliting. Conversely, many strategies revealed by the ThBA have little applicability to buildings simply because they involve movement and most buildings are static. During the evaluation of the ThBA by the panel of biologists, it was suggested that buildings might be more like plants than animals (see Chapter 8, Section 8.2), although thermoregulation in plants is complex, as the ThBA shows. Perhaps what needs to be remembered is that in many cases, it is the people inside buildings that need to be comfortable and people are part of the animal kingdom. The exceptions are buildings that need to be maintained at a particular temperature or within a particular temperature range because of what they contain, such as a cold store for fruit, or a hothouse for growing salad vegetables. If people are capable of thermoregulation then maybe more use should be made of this ability in the way buildings are run and organized. Suitable clothing for the climate in place of the office business suit is a simple and cheap strategy and in a recent study, adjusting clothing to keep cool was a much more frequent choice than turning on the ceiling fan or the air conditioning (Meinke et al. 2017). Schools that use passive or active systems of solar heating in colder climates are an example of running a building in such a way as to maximise what the climate offers, as the hours of building occupancy correspond with the hours when the sun is available. An example is the Maosi Ecological Demonstration Primary School in a village in China (CUHK 2008) which uses passive solar principles.

Clustering has already been mentioned as a strategy used by people in cold climates, as more people in one room can help to keep the temperature up. When it comes to hot climates, shading systems have been used for centuries as a means of keeping the sun out of the building interior. Perhaps more importantly, it seems such a simple principle has been forgotten in many modern glass buildings in which overheating of the zones in the solar-facing elevations leads to an increased use of air conditioning. What can be learned from thermoregulation strategies in the natural world is no animal would deliberately put themselves in such a situation. Each nest, and a building is no more than a human nest, is made appropriate for its purpose and climate.

9.2.1 Possible links revealed by the ThBA

While for most solutions in the final version of the ThBA (version 04), the parameters of the heat transfer equations for thermal regulation in architectural design are similar to their biological equivalents, there are some that are different. The reason for the difference appeared to relate to the limited number of parameters with the potential to be applied to building design. For a building, the thermoregulatory design opportunities are limited to the characteristics of the envelope, the spatial organisation and the design of the HVAC systems. This is because heat is gained either through the building envelope or via the internal gains of the occupants, equipment and plant, and heat loss is normally through the building envelope. The analogies to these would be the skin of animals or tissues of plants, the clustering behaviour of animals and respiratory and circulatory mechanisms (Table 9-1).

What Table 9-1 reveals is that designers are already making use of the principles of thermoregulation found in nature. This is hardly surprising

Table 9-1. Analogies between architecture and biology.

Where is the central thermoregulatory principle centred?	Heat	Biology	Architecture
Inside	Generation	Thermogenesis	HVAC and internal gains
Outside	Generation	Clustering	Space organisation
Inside	Transfer	Respiratory mechanism	HVAC (Air conditioning)
Inside	Transfer	Circulatory mechanism	HVAC (Cooling and heating pipes)
Outside	Transfer	Body	Envelope (glazing and opaque surfaces)
Outside and inside	Transfer	Cell, tissue, skin	Interior and exterior materials
Outside	Transfer	Between the organism and the environment	Between the building and the environment

since people are a part of nature and not separated from it. What is perhaps more intriguing is that the architectural translations of biological systems are generally achieved through morphological configurations while thermoregulation is a process. An example would be the Eastgate Centre in Harare which was an office building designed to have the characteristics of a termite's nest by having a vertical format with a chimney to exhaust the hot air. Unlike the termite mound which uses stack pressure and wind pressure to move the air, the Eastgate Centre relied on fans, which take energy for their operation. This suggests that the ThBA might be useful in making architects and designers think more about the process rather than just copying form.

9.3 Does nature hold the answer?

During the discussion with the panel of biologists who validated the inclusiveness of the ThBA (see Chapter 8, Section 8.2), the comment was made that nature is not always efficient, so perhaps looking at thermoregulation in natural organisms might not be the best place to start looking for inspiration for making energy-efficient buildings. The hippopotamus is claimed to be the most efficient mammal with "…a lifestyle which is energy efficient" (Timbuka 2012). They are vegetarian and eat 40 kg of grass each day, which is 1–1.5% of their body weight; in comparison, a cow on grass pasture will eat 2.5% of its body weight daily (Selk 2016). They also come inland from the rivers to graze just before nightfall, feed for up to five hours and return before dawn (Timbuka 2012), thus avoiding the heat of the day. Such efficiency as found in a hippopotamus is not immediately translatable to architecture, apart from noting that the hippopotamus is active at a time that least stresses the animal. The human comparison here is with the afternoon siesta that formed part of many societies in the past and that has continued in Spain until relatively recently (Jones 2017). Cao and Edery (2017) note that "… many diurnal animals exhibit a mid-day siesta that is more prominent at higher temperatures, almost certainly a critical adaptive response to avoid the detrimental effects of unnecessary exposure to heat." This is a thermoregulatory strategy that humanity could follow by adjusting the hours of the working day and that would have an effect on buildings, either through providing space in workplaces for the siesta or through ensuring workplaces were within easy walking or cycling distance of homes. The problem is that although such changes would be inspired by the natural world, they would also require fundamental changes in society, and so are outside the remit of normal designers.

Nieuwenhuis (2016) makes that point that nature is resilient rather than efficient. "Unlike humans, nature is more resilient but far less efficient: numerous plant seeds are dispersed just to allow some to germinate, and many animals have extremely short lifespans—both suggest a wasteful use of resources." What nature does is to be effective rather than efficient, a concept which might be applicable to buildings. When applied to manpower, efficiency means doing a task in the shortest time while effectiveness is more to do with the quality of the output from doing the task. On the whole, it might be better to take longer and produce a more durable and satisfactory outcome.

Building designers are familiar with the concept of cost effectiveness, which is trying to get the best building for the money available, in other words the best outcome. McCardell (2018) defines building energy effectiveness as a building-based organization using "…energy conservation, energy efficiency, water use reduction, and renewable energy systems and approaches in ways which allow the organization to better accomplish its objectives." Effectiveness also recognises that things change, as does the concept of resilience in nature. A building designed to be energy efficient today may not work as well if the workforce is suddenly expanded within the same building envelope. Other things may change, such as the type of energy used by a building as countries move to decarbonising energy supplies. The problem is the designer tends to be designing an energy-efficient building by guessing what the future parameters might be. Is it better to invest in more insulation now, or to invest in renewable energy generation on site? This introduces the idea of trade-offs, something that is also found in the natural world.

9.3.1 Trade-offs

Trade-offs are common in the natural world. If food is short then birds might not breed but wait to the next season (Sumasgutner et al. 2014), although lack of food is not the only parameter that might limit bird populations. Mention has already been made of trade-offs when it comes to thermoregulation. It takes energy to dig a burrow in order to retreat from the heat of the day, which means finding more food to produce that energy, which requires moving to find the food. Similar trade-offs are emerging in the field of energy-efficient buildings. Increasing the insulation, and possibly mass, in a building will increase the energy embodied in the construction, so care has to be taken that over the life of the building, the savings in operating energy will outweigh the energy embodied in the additional materials that make these savings possible. The same is true of measures such as shading devices for reducing cooling loads. Some type of energy balance investigation is necessary over the predicted life of the building, including materials that are needed for maintenance and refurbishment of any energy saving devices. Inspiration from nature is not enough on its own without ensuring that it does lead to a reduction in building energy use.

9.3.2 *Unknown nature*

It might be that nature has secrets about thermoregulation that are yet to be revealed and which might be of use for those designing buildings. For example, the concepts of endothermy and ectothermy are thought to be more complex than research has so far revealed. Legendre and Davesne (2020) make the following comment:

> "… endothermy in mammals and birds is not as well defined as commonly assumed by evolutionary biologists and consists of a vast array of physiological strategies, many of which are currently unknown. We also describe strategies found in other vertebrates, which may not always be considered endothermy, but nonetheless correspond to a process of active thermogenesis."

The same study also notes that the physiological processes behind heat production for mammals and birds are very costly, which relates back to the idea of trade-offs described above. Perhaps the most useful idea to emerge from this research is that thermoregulatory strategies are often used in combinations, and that doing this is fundamental to their survival. The researchers also provided a comprehensive summary of the various ways vertebrates show endothermic characteristics together with the various anatomical specifications that enable thermal adaptation. For example, heat generation could take place through shivering, non-shivering, red muscle, cranial or a combination of these.

It also seems that the hierarchical connections in plants could be important when it comes to designing a new generation of biomimetic technologies. An example is the development to proof of concept stage of innovative PVs, designed to perform efficiently in low light environments. To do this, researchers studied different aspects of light regulation in plants at the organ, tissue and cellular levels (Yun et al. 2019). While this is not related to thermal management, it highlights the importance of investigating the interconnectedness of the mechanisms plants use to respond to environmental stimuli.

9.4 Waiting for new technology

Using the ThBA and looking for parallels in architecture for biological thermal adaptation strategies revealed that a large number of biological solutions are currently employed in building design. However, for most of the architectural equivalents, the biological solutions represent at best a simple translation of the intricate natural application of their heat transfer principle. This means the sophistication and internal hierarchical connections of almost all physiological strategies remain unexplored. The main barrier to their exploration is that the technology falls short when compared to the hierarchically organised, dynamic and multiscale operational characteristics of living things (cells, tissue, organs). The following summarises the design opportunities that biological organisms could provide for future buildings:

a) **Circulatory mechanisms:** In vasoconstriction, thermal regulation happens through heat transfer between blood vessels and skin, these being the internal and peripheral tissues. Consider blood vessels as pipes with a fluid flowing through them that carries the thermal energy to where it is needed in a building; the smaller the diameter of the pipes, the less fluid will flow, and thus, the less heat will be lost through conduction and convection during the energy transfer. The big difference between blood vessels and current ducts found in HVAC systems is that the latter never change in size in response to changing external stimuli. A bio-inspired design solution might be creating a mesh of capillary pipes that can change their diameter and that are embedded between the materials used for either external or internal surfaces. Having flexibility in narrowing the fluid channels should decrease heat loss due to less fluid flowing in the pipes. However, new flexible materials that could do this would have to be developed. At present, the simple solution borrowed from the natural world is to insulate the pipes.

b) **Temporary solutions necessitate dynamic movement:** As nearly all biological thermal adaptation solutions are reversible, this poses a potential problem for their application in buildings, as the equivalent solutions would need to be alterable to allow them to be exact analogies of their biological parallels. For example, altering the material properties of the building envelope to respond to changes in the external conditions that are reversible is problematic. The 1987 Arab World Institute in Paris by architect Jean Nouvel had a moveable façade that responded to light, just as a flower might open in response to light. "The 240 motor-controlled apertures consist of 113 photosensitive panels with 16000 moving parts and 30000 light-sensitive diaphragms" (McKiernan 2013). Sadly, the technology did not match the design and after a while the façade failed to work as intended.

- Like burrows, buildings have been buried and semi-buried in the ground to take advantage of the reduction in temperature swing this brings, such as the UK Hockerton Housing Project (Annable 2006) (see Figure 8-5). The next generation of buildings might be capable of being buried in the ground during extreme periods of the year and emerge back on the surface when such conditions are over. They might be able to move on their construction site shuttling between sun and shade. The movement of the whole building to a cooler environment is expected to be more effective than shading devices, as the latter only keep part of the envelope cool.

Animals migrate to avoid harsh climates for a certain period of time, as do some wealthy people who move to warmer climates in the winter but it is unlikely that buildings will ever migrate. It might be possible to create a microclimate around a building to mediate between an unwelcome outside temperature and a more ideal interior one, but this is not analogous to migration.

- Dynamic changes in building geometry and size, the orientation and size of shading devices and space organization all have the potential to imitate temporary thermoregulatory solutions in nature. However, the technology to do these things in a way that saves energy without adding energy to the buildings is still to be explored in detail.

c) **Permanent solutions:** The ThBA revealed several permanent solutions for different actions that could perhaps be translated to architecture. The permanent solutions used by animals were insulation, countercurrent heat exchange and the orientation and material properties of structures like nests (birds) and mounds (insects). Most insulation in buildings is permanent, but insulated shutters have been used on windows in cold climates as a means of reducing heat loss, so here the insulation level is changeable as the shutters are open to let light and sunlight in during the warmer day.

- The principle of countercurrent heat exchange could be used to transfer heat from a hot zone to a cooler zone in a building.

- The optimal choices when it comes to orientation and materials are as crucial to making an energy-efficient building in any climate as they are to making a satisfactory nest.

d) **Finding inspiration from thermoregulatory mechanisms happening at the organ, tissue and cellular levels:** The thermal regulation principles behind some active solutions in plants occur at the cellular and molecular level through biochemical reactions. These mechanisms could be useful for building design as morphological changes in cellular levels could be translated to the design of the envelope using the nanostructure of materials.

e) **Buildings are not living organisms:** It seems buildings cannot be called 'living organisms', although in their 1960 manifesto (Metabolism: Proposals for a New Urbanism), the Japanese Metabolists "used biological metaphors to call for buildings capable of regeneration" (Cohen 2019), regeneration being a characteristic of living things. Even the phrase 'living architecture' (Eng et al. 2001, Garnier et al. 2013, Flynn 2016) does not seem to do justice to what 'living' literally means. Apart from behavioural thermal adaptation, the two physiological thermoregulatory solutions in plants and animals and some morphological thermoregulatory solutions that only occur in plants, happen in a hierarchical way, meaning that a specific change in a chemical reaction has an impact on the cellular level, which would then cause change in a tissue to enable thermoregulation. Envisaging a building that could do this does not seem feasible given the current state of architectural technology. For this to happen, HVAC equipment and layouts, space layouts, the materials used for interior and exterior surfaces and the size and shape of glazing and opaque surfaces would need to be synchronised to allow a dynamic and multilevel thermoregulation similar to biological thermoregulation to happen. The only exception are the insulating tissues used by organisms to control temperature and airflow. These are counted as exceptions when it comes to physiological solutions, as their translation into insulating materials for buildings is already a well-developed technique.

Unlike physiological and morphological adaptations, the behavioural thermal adaptation mechanisms used by organisms have been translated to architectural design, since almost all animal behaviours except the three types of torpor, are independent of the internal thermoregulatory principles. For these behaviours, interaction with the environment is the most important principle in thermoregulation. Animals might move in space, orient themselves towards the sun and change shape, colour and posture; all these ideas have a place in building design. However, the building has to be made to move through human intervention, whether physical or through some kind of integrated energy system designed by a person. Together, this suggests that envisaging buildings as living organisms is not possible, and remains in the world of poetic analogy.

f) **Buildings tend to keep operating:** Referring back to the definitions for daily torpor, estivation and hibernation (see 6.2.3, "avoiding heat loss") an equivalent strategy in the built environment would be to shut the building, such as an office, down in winter and to ask the occupants to come back to it in summer. However, as noted earlier, some buildings like schools do shut down periodically, though perversely in cold climates, schools run during the winter, and so need heating, and shut in the warmest summer months for the holidays. Some structures, like unheated conservatories attached to a house, might only be used on warm days but then having a structure that is not often used may, of itself, be energy inefficient.

g) **Kinetic movements might not be efficient:** As mentioned earlier, the static structure of buildings seems to be a barrier to them using the thermoregulatory strategies of living organisms. There have been suggestions for creating kinetic architecture that could imitate the movement capability of organisms through partial movements of a building's structure (Bayhan and Karaca 2019). The adjustable and dynamic character of kinetic design follows the rhythms of nature (El Razaz 2010), but it still needs to be evaluated from an energy efficiency point of view, as energy balance calculations are required to know whether the bio-inspired design approach is more efficient in overall energy terms.

9.5 What was learned from developing the ThBA

Looking back on the development of the ThBA, the following points seem worthy of consideration:

- When comparing thermal adaption mechanisms in nature to architecture, it seems organisms use a tight feedback loop in response to thermal challenges. This means, where organisms cannot adapt their body temperature to the environment through using a certain thermal adaptation mechanism, they intelligently switch into another one and this keeps repeating until thermal comfort is achieved. However, in buildings, the same series of ideas might not be possible. A poor analogy in the context of thermal control of buildings, is the use of a thermostat. However, the on/off switching behaviour of a thermostat is not the same as the biological switching patterns that intelligently enable transition from one strategy into another.

- It seems that thermal adaptation strategies do not vary from species to species or, in other words, there is not an infinite number of adaptation strategies in nature. This, accordingly, might suggest the irrelevancy of assuming nature to be the infinite source of intelligent ideas, at least in the context of thermal adaptation. Comparing the architectural parallels to biological ones might also reveal gaps where buildings have fallen short of emulation or at least proper imitation of the thermoregulatory strategies used by natural organisms.

- The ability of organisms to respond to thermal stresses by using a hierarchy of thermal adaptation methods does not seem to have a parallel in buildings. While thermal adaptation in buildings can also be achieved through using a hierarchy of strategies such as openable windows, insulation and sunshades, this is not comparable to biological systems in which there is a smart and automated connection between the different levels of the hierarchy.

What has to be remembered is that people are part of the natural world and that buildings are themselves a human thermoregulatory strategy the makes living in the varied climates colonized by human beings possible. Many vernacular buildings used strategies that had immediate parallels with the way animals and plants used thermoregulation in the same environments. The problem is that there are now too many people on a small planet, and to make life tolerable, energy has to be used in buildings that are very different from their vernacular precedents. There are known ways of making these buildings more energy efficient without having to look to the natural world for ways of doing this. What nature teaches is that you have to act to survive rather than wait for a new genius idea, whether bio-inspired or not, before you act.

References

Annable, R. (2006). Hockerton Housing Project—Autonomous housing in Nottingham, UK. Retrieved 24 March 2019, from Flickr.com https://www.flickr.com/photos/eversion/300040169/in/album-72157594381130837/.

Badarnah, L. (2012). *Towards the Living Envelope: Biomimetics for Building Envelope Adaptation.* (Doctoral dissertation), TU Delft, Zutphen, Netherlands.

Bayhan, H. G. and Karaca, E. (2019). *SWOT Analysis of Biomimicry for Sustainable Buildings—A Literature Review of the Importance of Kinetic Architecture Applications in Sustainable Construction Projects.* Paper presented at the World Multidisciplinary Civil Engineering-Architecture-Urban Planning Symposium (WMCAUS) 2018, Prague, Czech Republic.

Cao, W. and Edery, I. (2017). Mid-day siesta in natural populations of D. melanogaster from Africa exhibits an altitudinal cline and is regulated by splicing of a thermosensitive intron in the period clock gene. *BMC Evolutionary Biology, 17*(1), 1–17.

Cohen, A. (2019). The Japanese Architects Who Treated Buildings like Living Organisms. Retrieved from https://www.artsy.net/article/artsy-editorial-japanese-architects-treated-buildings-living-organisms.

CUHK. (2008). CUHK Award-winning Ecological School Sets Model for Future Buildings in Rural China. Retrieved from http://www.cuhk.edu.hk/cpr/pressrelease/081124e.htm.

El Razaz, Z. (2010). Sustainable vision of kinetic architecture. *Journal of Building Appraisal, 5*(4), 341–356.

Eng, K., Bäbler, A., Bernardet, U., Blanchard, M., Briska, A., Costa, M., Delbrück, T., Douglas, R., Hepp, K., Klein D., Manzolli, J., Mintz, M., Netter, T., Roth, F., Wassermann, K., Whatley, A., Wittman, A. and Verschure, P. (2001). *Ada: Buildings as Organisms.* Paper presented at the Game, Set, and Match Symposium, Delft, Netherlands.

Flynn, E. (2016). (Experimenting with) Living architecture: a practice perspective. *arq: Architectural Research Quarterly, 20*(1), 20–28.

Garnier, S., Murphy, T., Lutz, M., Hurme, E., Leblanc, S. and Couzin, I. D. (2013). Stability and responsiveness in a self-organized living architecture. *PLOS Computational Biology, 9*(3), e1002984.

IEA. (2019). Energy Efficiency: Buildings, The global exchange for energy efficiency policies, data and analysis. Retrieved from https://www.iea.org/topics/energyefficiency/buildings/.

Jones, J. (2017). It's time to put the tired Spanish siesta stereotype to bed. Retrieved from https://www.bbc.com/worklife/article/20170609-its-time-to-put-the-tired-spanish-siesta-stereotype-to-bed.

Legendre, L. J. and Davesne, D. (2020). The evolution of mechanisms involved in vertebrate endothermy. *Philosophical Transactions of the Royal Society B, 375*(1793), 20190136.

McCardell, S. (2018). *Energy Effectiveness: Strategic Objectives, Energy and Water at the Heart of Enterprise*. Springer International Publishing AG, part of Springer Nature. e-Book.

McKiernan, M. (2013). Jean Nouvel, Arab World Institute, 1987. *Occupational Medicine, 63*(8), 524–525.

Meinke, A., Hawighorst, M., Wagner, A., Trojan, J. and Schweiker, M. (2017). Comfort-related feedforward information: occupants' choice of cooling strategy and perceived comfort. *Building Research & Information, 45*(1-2), 222–238.

Nieuwenhuis, P. (2016). Humans strive for efficiency but could learn so much from nature's resilience. Retrieved from https://theconversation.com/humans-strive-for-efficiency-but-could-learn-so-much-from-natures-resilience-66103.

Selk, G. (2016). How Much Hay Will a Cow Consume in a Day? Retrieved from https://nwdistrict.ifas.ufl.edu/phag/2016/11/18/how-much-hay-will-a-cow-consume-in-a-day/.

Sumasgutner, P., Nemeth, E., Tebb, G., Krenn, H. W. and Gamauf, A. (2014). Hard times in the city–attractive nest sites but insufficient food supply lead to low reproduction rates in a bird of prey. *Frontiers in Zoology, 11*(1), 1–14.

Timbuka, C. (2012). *The Ecology and Behaviour of the Common hippopotamus, Hippopotamus amphibious L. in Katavi National Park, Tanzania: Responses to Varying Water Resources*. (PhD Thesis), University of East Anglia.

Yun, M. J., Sim, Y. H., Cha, S. I. and Lee, D. Y. (2019). Leaf anatomy and 3-D structure mimic to solar cells with light trapping and 3-D arrayed submodule for enhanced electricity production. *Scientific Reports, 9*(1), 1–9.

Index